CULTIVATING BIODIVERSITY

CULTIVATING BIODIVERSITY

Understanding, Analysing and Using Agrodiversity

Harold Brookfield, Christine Padoch, Helen Parsons
and Michael Stocking

Practical
ACTION
PUBLISHING

published in association with

UNITED NATIONS
UNIVERSITY

UNEP

Practical Action Publishing Ltd
25 Albert Street, Rugby, CV21 2SD, Warwickshire, UK
www.practicalactionpublishing.com

© The United Nations University 2002
First published 2002

ISBN 1 85339 493 9
ISBN 13 Paperback: 9781853394935
ISBN Library Ebook: 9781780441092
Book DOI: https://dx.doi.org/10.3362/9781780441092

A catalogue record for this book is available from the British Library.

The authors, contributors and/or editors have asserted their rights under
the Copyright Designs and Patents Act 1988 to be identified as authors of
their respective contributions.

Since 1974, Practical Action Publishing has published and disseminated
books and information in support of international development work
throughout the world. Practical Action Publishing is a trading name
of Practical Action Publishing Ltd (Company Reg. No. 01159018), the
wholly owned publishing company of Practical Action. Practical Action
Publishing trades only in support of its parent charity objectives and any
profits are covenanted back to Practical Action (Charity Reg. No. 247257,
Group VAT Registration No. 880 9924 76).

Index prepared by Indexing Specialists (UK) Limited
Typeset by Dorwyn Ltd, Rowlands Castle

Contents

Part III. Agrodiversity case studies

Part IV. The way forward

Introductory note and acknowledgements

HAROLD BROOKFIELD

THE INTRODUCTION TO this book is in Chapter 1. The book arises out of the United Nations University project on People, Land Management and Environmental Change (PLEC), and all its authors are members of that project. More than half of the chapters are based on material that has appeared previously in the project's periodical, *PLEC News and Views*, which began to appear about half-yearly in 1993 and has continued into 2002. A few are reprinted from that periodical almost unchanged. *Cultivating Biodiversity* now offers this material wider dissemination. The other chapters are either newly written for this book, or are so much changed from earlier versions in *PLEC News and Views* as to be, in effect, new writing.

This note is principally to express our acknowledgements, first and foremost to the world's small farmers who cultivate biodiversity. The present generation represents all those millions who have evolved diverse methods of production in many lands over many centuries. Among the skilful present-day farmers, we offer our special thanks to the expert farmers in the PLEC demonstration site areas who have worked with us and have taught us most of what is to be found in this book.

We acknowledge the support of the United Nations University (UNU), and the Global Environmental Facility, which has made this work possible. In particular, within UNU we are grateful to Juha Uitto and Audrey Yuse for the enormous amount of support and encouragement they have given us over the years, and to Liang Luohui, who became our managing coordinator in 1998. Liang also appears as an author in this book. The editors and other authors are also grateful to Masako Ebisawa, who has taken Audrey's place as our financial overseer. Among our international supporters, we are particularly grateful to Timo Maukonen of UNEP and Chona Cruz in the Global Environmental Facility for having worked so hard on our behalf at several critical times.

The editors greatly appreciate the support of their institutions: the Australian National University, and especially its Department of Anthropology; the New York Botanical Garden; and the University of East Anglia. In Canberra we gratefully acknowledge the skilful work of Kay Dancey, who has redrawn or amended several of the maps. In our own small circle, we are grateful to Muriel Brookfield and Ann Howarth for their constant support, and for their dedicated work on the 12 papers that were initially published, in the same or different form, in *PLEC News and Views*. In the larger world beyond, we are grateful to the institutions of all our authors, and to their colleagues who have worked without reward to ensure the success of the PLEC enterprise.

Figures

Tables

Boxes

Acronyms and abbreviations

ABA	Agrobiodiversity assessment
CAMPFOR	Compania Amazônica de Producción Forestal
CBD	Convention on Biological Diversity
CBO	Community-based organization
CCD	Convention to Combat Desertification
CGIAR	Consultative Group on International Agricultural Research
COP	Conference of Parties
DBH	Diameter at breast height
FAO	Food and Agriculture Organization (of the United Nations)
GEF	Global Environment Facility
GIS	Geographic information system
HH-ABA	Household level agrobiodiversity assessment
HYV	High yielding variety
IARC	International Agricultural Research Centre
IBSRAM	International Board for Soil Research and Management
ICRAF	International Centre for Research in Agroforestry
IPGRI	International Plant Genetic Resources Institute
NGO	Non-governmental organization
PCA	Principal component analysis
PLEC	People, Land Management and Environmental Change
PLEC-BAG	PLEC Biodiversity Advisory Group
SRL	Sustainable rural livelihoods
TSBF	Tropical Soil Biology and Fertility Programme
UNDP	United Nations Development Programme
UNEP	United Nations Environment Programme
UNESCO	United Nations Educational, Scientific and Cultural Organization
UNFDAC	United Nations Fund for Drug Abuse Control
UNU	United Nations University

Part I. Understanding and analysing agrodiversity

1 Cultivating biodiversity: setting the scene

THE EDITORS

THE GREAT ADVANCES in agricultural production of the past 50 years have not been achieved by increasing the diversity of crops and farms, but rather by decreasing the diversity. It is a fact that most growth in production has been attained on every continent by substituting one or two high-yielding varieties (HYVs) or hybrids for the traditional abundance of species and varieties, by using chemicals to wipe out diverse communities of insects and other pests, by substituting a single prescribed management routine for a profusion of cropping methods, and by replacing a hundred communities of farming families with a few large, highly mechanized enterprises. Other land uses have similarly seen a replacement of complex vegetation associations by uniform stands and consistent practices. Forest production is now dominated by rows of single species, usually exotic. The requirements of irrigation mean single crops and standard practices. Mechanization has encouraged larger fields, the displacement of hedgerows and the use of artificial fertilizers. Such is the history of agricultural change in the modern era; but it is not the whole history.

While multiple 'Green Revolutions' have changed the landscapes of many regions, the HYVs have made few inroads in others. These are areas where the miracle rices, wheats and cash crops do not dominate. They may thrive for a short while, but the new varieties then take their place alongside traditional varieties, or disappear entirely. In some regions periodic droughts eliminate those water-hungry, high-yielding varieties that farmers try to plant; in others, excess flood waters drown or sweep away crops unadapted to periodic inundation. A lack of chemical inputs or irrigation makes the planting of demanding crops impossible in many remote and resource-poor regions. Poor transport, marketing and storage make reliance on one or two varieties far too risky for the rural poor. With the recent reduction in the subsidies that have been available since the 1960s, many poor farmers can no longer afford to buy seed of the HYVs. It is true that there are some areas where the new crops have proved profitable and are produced in large quantities for markets. But among these scientific advances, traditional crops, unusual varieties, non-standard methods, and innovative approaches still flourish.

Throughout the world some farmers cannot or will not give up their own agricultural choices in favour of a wholesale substitution of the new. They value their tasty, beautiful, hardy, historic, or holy plants, and steadfastly hold on to a variety of landraces and diverse cropping methods. There are many forward-looking and profit-seeking farmers in these 'marginal' areas who have experimented with many crops and varieties and have found the modern packages lacking. Based on a profound knowledge of both their natural and social environments and an appreciation of fickle markets, they have developed new varieties. They have preserved or improved old varieties, discovered how to control a devastating crop disease by increasing shade and plant diversity in their fields, designed new ways to prepare the soil in order to prevent water erosion, or have overcome land or labour shortages by creating multilayered and multi-species agroforests. They have learned how to spread the risk of farming in difficult environments by diversifying their use of whole landscapes, and mixing their own agricultural practice with off-farm activities. These exceptional men and women have combined much of the best of the old and the new, the local and the borrowed, to come up with ever-evolving ways of making a living in tough and dynamic situations.

We call what they do agrodiversity, meaning the many ways in which farmers use the natural diversity of the environment for production. This includes not only their choice of crops but also their management of land, water and biota as a whole. The biodiversity of crop plants, agricultural biodiversity or agrobiodiversity is a subset of this agrodiversity, now an important field of study in its own right, and the relation between the two is discussed more fully in Chapter 2. This book is about agrodiversity, not only as practised by the best farmers, but also over large areas and by whole societies.

This introductory chapter sets the stage for *Cultivating Biodiversity*. Our title proposes that the essentially natural virtues of biodiversity can be managed successfully for human purposes. An analogue of natural biodiversity can be created on a farm. It is a different biodiversity, but nonetheless a valuable one. It may have many of the presumed benefits of, say, a multispecies forest in providing habitats for other species and in conserving varieties and genes that society may come to want in the future. However, it will be managed largely for utilitarian purposes. Agrodiversity, as featured in the subtitle of this book, implies that this can be done as part of productive land use. Much of this book is given over to demonstrating, from the People, Land Management and Environmental Change (PLEC) project network, examples of good practice in agrodiversity, where environmental and human developmental goals coincide. Without such demonstrations, *Cultivating Biodiversity* will remain empty rhetoric – good in theory, but poor in practice.

The modern decline in diversity

Jules Pretty (1995) estimated that some 75 per cent of the genetic diversity of agricultural crops has been lost in the past 100 years. Only about 150 plant species are now commonly cultivated for food, and just three of these supply nearly 60 per cent of calories derived from plants (Fowler and Mooney, 1990). For just one of these three crops, rice, India used to have some 30000 varieties; now only ten provide for over three-quarters of production. Loss in species and varieties is mirrored in loss of a diversity of traditional techniques in cultivation. Observant colonial agriculturists such as N.V. Rounce (1949) documented some of the complex techniques of mulching, soil mounding and land preparation techniques of the WaSukuma of Lake Province of Tanganyika. Rounce was in the vanguard of promoting 'modern' techniques, and was dismissive of the skills of the African cultivator ('the African is not entirely unaware of the value of periods of rest under grass, weeds or bush', p.40). Even so, he described many practices that would be lauded today. Until, for Africa, Allan (1965) started to bring knowledge together, much agro-diversity must have gone unnoticed and undocumented.

Modern science fostered much of the disappearance of diversity. Pretty (1995: 76, quoting Stakman et al., 1967) relates the visit of the Rockefeller-funded team to Mexico in the 1940s to assess wheat cultivation before recommending a national breeding programme. Dismissive of low-yielding, traditional fields, they wrote:

> most fields were a mix of many different types, tall and short, bearded and beardless, early ripening and late ripening; fields usually ripened so unevenly that it was impossible to harvest them at one time without losing too much.

Crop breeding and promotion campaigns, as well as demonstrated increases in yield, spurred much of the loss of varietal diversity. Species diversity declined largely because marketing and incentive structures failed to provide opportunities to grow and sell traditional crops. Cheap and readily available inputs meant that farmers specialized in only a few crops. In a few cases, farmers were forced by law or the denial of subsidies to simplify their planting. Similarly, the rise of agricultural mechanization meant standardization of soil preparation techniques, supported by extension bulletins on exactly how and when to plough, and land use directives which dictated what practices should be undertaken on which slopes and by what techniques. A complex of economic changes in many countries set millions of smallholding families on the road to the cities to work in factories and offices, their creativity and expertise lost to agriculture.

Agricultural research even today continues the generally dismissive view of traditional, biodiverse agriculture and its possible role in development. Pretty (1995: 6) has assembled a 'box' of quotations from the well-known 'actors' in modernist perspectives on agriculture. Each actor plays a well-observed stereotypical role: for example, an influential consultant to the CGIAR and World Bank tells leaders of developing countries not to be duped into believing that future food requirements can be met through 'impractical low-input, low-output technologies'. Another actor reminds us that, 'the adoption of science-based agricultural technologies is crucial to slowing – and even reversing – Africa's environmental meltdown'. Such alarmist predictions, intended to promote the role of science, have pervaded most writings on environmental change, especially for Africa (Stocking, 1996). The effect, intended or not, has been to decrease biodiversity, and threatens to reduce agrodiversity to a number of isolated pockets where modernization is either resisted or has failed to arrive. Fortunately, this has not yet occurred. Agrodiversity, as Brookfield (2001: 285–6) has argued, is inventive and adaptable and, for this reason, has the resilience that has enabled it to survive, in spite of all pressures, right through the twentieth century.

Biodiversity as part of the global environmental change agenda

Since the 1980s, the global environmental change agenda has seen to it that biodiversity has been promoted as a good thing in itself. Much conservation effort has been targeted at protecting biodiversity on the level of genes, species, habitats and ecosystems. There are important germplasm collections for most major food crops. A lot of the effort has gone into designating protected areas, in which human impacts are excluded or minimized. By the mid-1990s, nearly 6 per cent of the earth's land surface had been designated as protected, with 3.5 per cent in scientific reserves or national parks (Primack, 1998). Such efforts have saved some rare varieties and threatened habitats, but only through what is often called 'fortress conservation' – expensive germplasm storage facilities and areas strictly protected from human land use. Biosphere reserves, promoted by the UNESCO Man and Biosphere programme, are an alternative model intended to demonstrate that conservation could be compatible with sustainable development. By 2001, 393 reserves had been created in 94 countries, covering more than 3 000 000 hectares. More recently, various forms of integrated conservation and development projects have proliferated, demonstrating how protected areas can also allow sustainable resource use by local people. Many countries, notably Brazil, have created local innovative designs that attempt to combine conservation with the satisfaction of the needs of rural people. These designs have greatly varying track records; many have shown that the costs of conserving biodiversity can be extremely high, bringing into question how difficult it would be to replicate these experiences even more widely.

At a global level, it was the Convention on Biological Diversity (CBD), introduced at the Earth Summit in Rio de Janeiro in 1992, that has done most

to promote the notion that biodiversity in areas of land use is worth protecting. The CBD entered into force in 1993 and by March 1999 had 174
signatories, only two fewer than the most ratified convention (Climate
Change), but nonetheless with some notable absentees (USA and Thailand,
for example). Agrobiodiversity features prominently in the provisions of the
CBD. The Third Conference of Parties to the Convention, meeting in Buenos
Aires in 1996, recommended:

- identification of key components of biological diversity in agricultural production systems
- implementation of targeted incentive measures which have positive
  impacts on agrobiodiversity, in order to enhance sustainable agriculture
- encouragement to develop technologies and practices that increase productivity, arrest degradation, reclaim land and restore biological diversity
- empowerment of indigenous and local communities to build capacity for *in
  situ* conservation and sustainable use of agricultural biodiversity
- promotion of partnerships between researchers, extension workers and
  farmers for agrobiodiversity
- promotion of Integrated Pest Management, particularly methods which
  maintain biodiversity
- studying the positive and negative impacts of intensification on ecosystems
  and biomes.

These were landmark goals for the international conservation community.
They moved conservation efforts from the purely ecological sphere and
placed agrobiodiversity firmly on the agenda of sustainable development.
Quite contrary to the situation of a decade earlier, many authors now see
agricultural biodiversity as the most important part of biodiversity for human
survival (e.g. Wood and Lenné, 1999). Within this general framework for sustainable development, global strategies of biodiversity conservation would
thenceforth have to fulfil critical needs for food security, farm productivity,
social equity and welfare, and the rights of local people as well as ecological
integrity (Thrupp, 1998).

The funding mechanism to assist developing countries in fulfilling their
obligations under the CBD is known as the Global Environment Facility
(GEF). The content of the GEF 'Operational Programs' are good indicators
as to how far reality has caught up with rhetoric in recognizing the key
importance of land use in conserving biodiversity. GEF Operational
Program no. 13 (GEF, 2000) was written to elaborate the recommendations
of the Fifth Conference of Parties (COP5), meeting in Nairobi in 2000, that
agricultural biodiversity must be given extremely high priority. COP5
identified four main programme elements that should be taken on board by
the GEF:

○ assessments of the status and trends of the world's agricultural biodiversity, as well as analysis of causes of change
○ adaptive management to identify practices, technologies and policies that would promote the positive and mitigate the negative impacts of agriculture on biodiversity
○ capacity building to strengthen the ability of farmers, indigenous and local communities to manage agricultural biodiversity sustainably
○ 'mainstreaming' to support national plans and strategies for the conservation and sustainable use of agricultural biodiversity.

The new GEF Operational Program lists 14 types of activity that it is prepared directly to fund in support of agricultural biodiversity, several of which include aspects of agrodiversity as outlined in this book.

About PLEC

Against this background of now-substantial international support for agrodiversity, one GEF-funded project stands out as a pioneer – and this is the project on which this book is based: People, Land Management and Environmental Change (PLEC). PLEC has been investigating agrodiversity, and working with the farmers who practise it, since the early 1990s. The United Nations University started the project in 1992. By 2001, at the end of a four-year period of funding by the Global Environmental Facility, it had brought together the efforts of over 200 scientists, their students and technicians, and a much larger number of collaborating farmers in 12 countries across the developing world. At sites chosen by its members, PLEC focuses on the manner in which farmers successfully manage biodiversity, and make a good living while doing so. There are 21 developed demonstration sites across the world, and an additional eight or ten in which useful basic work has been done. Many of the demonstration sites in the eight countries where work is supported by GEF have farmers' associations, of one type or another, which have been formed by the farmers themselves to manage their work with PLEC support.

PLEC was set up between 1992 and 1994. Since then the project has been an association of locally-based Clusters of researchers and their collaborators in Brazil, China, Ghana, Guinée, Jamaica, Kenya, Mexico, Papua New Guinea, Peru, Tanzania, Thailand and Uganda. All groups are multi-disciplinary, but with differing disciplinary emphases. Ecologists, botanists, agricultural scientists, soil scientists, geographers and anthropologists are all strongly represented. All Clusters are based in national organizations or universities, involving collaboration usually between government, non-governmental, educational and scientific institutions. Coordination has been distributed between Tokyo, Canberra, New York and Norwich, and there has been no central project office. With the exception only of the managing coordinator in Tokyo, and some locally-recruited project assistants, all personnel have been, and remain, part-time. Collaborating scientists and farmers have given freely of their time to become involved in the goals of PLEC and its international network.

Initially, PLEC was a research project focusing on methods of farmers' management of their resources, especially (but not only) biodiversity resources. Before long, this research began to reveal that a minority of farmers, the ones we call expert farmers, managed their resources better than others, or at least in ways that seemed to gain greater and more assured production while at the same time conserving or creating biodiversity and reducing environmental degradation. Our research sites then became the demonstration sites described above, where the teachers were increasingly the farmers rather than the scientists. Expert farmers may or may not cherish biodiversity *per se*, as professed environmentalists do. They do, however, know how to use diversity, and to appreciate its utility; they are not averse to the beauty of diversity, often mixing flowering and ornamental species with food crops around the homestead. They employ a variety of plants, animals, technologies and institutions to arrive at their production goals. In making agrodiversity useful by having it solve problems, the expert farmers provide an important key to preserving, producing and passing it on.

Wherever social conditions make local farmer-to-farmer interaction feasible, PLEC has relied more and more on its expert farmers to advise others, with the scientists playing only a supporting role. The scientists do help directly by introducing new ideas, and also germplasm, from other places, but the method is quite different from standard extension method in which crops and culture packages are developed on the experiment stations, and then demonstrated to the farmers. Just how different the PLEC method is from the standard extension model comes through at many places in this book.

The evidence from most of the PLEC Clusters is that use of skilful and diverse methods enables farmers not only to sustain biodiversity and improve their soils, but also to increase production, while providing greater security of both food supply and incomes. Livelihoods can be secured at the local level, while at the same time global objectives of biodiversity conservation are met. Yet it continues to be argued widely that only energy-intensive monoculture can achieve the food production needed for a growing world population. To counter this argument, we present the fact that the working environments of most farmers practising agrodiversity are not only unsuitable for the use of such methods, but would be liable to serious land degradation if they were followed. Additionally, we argue from the evidence of the PLEC demonstration sites that agrodiverse practices are as productive, if not sometimes more so, than monocultures, and they meet human needs for variety and interest much more closely. Production, to be sustainable, requires that resources be conserved or enhanced. Moreover, as the evidence in this book shows, many farmers can do quite well while practising diverse conservationist methods of agriculture.

About this book

In this book, PLEC authors are speaking not only to their own colleagues, but also to many others who are interested in new approaches to rural development. PLEC's emphasis on working with farmers is not simply another version of 'participatory' research and action. Agrodiversity is not a technology that can be reduced to a simple formula, but there are lessons to be learned from PLEC's experience, and this book sets them out. Part I, concerned with the analysis of agrodiversity, includes this chapter and three other discussion chapters by two of the editors, followed by two which are based on 'how to do it' guidelines on data collection for project members, prepared toward the end of the project's first GEF-funded year. A further chapter introduces a variant approach to biodiversity assessment developed in China. The final chapter in Part I was written in 2001 to provide advice on quantitative analysis of agrodiversity for project members: it is included here because it has much wider potential utility, and offers a very clear statement on a topic that many find difficult.

In Part II, the book turns more specifically to PLEC's focus on the people. After an introductory analysis of the sometimes almost-invisible (to the outsider) nature of farmer expertise, we reproduce the major statement on demonstration site activities by members of the Demonstration Advisory Team formed in 1999. There follows a classic piece from PLEC's early days, on indigenous knowledge, by Kojo Amanor. Next is a revealing chapter on PLEC experiences with farmers in Tanzania. The section concludes with a statement on PLEC's most distinctive farmers' association, entirely female, at Jachie, Ghana.

Part III is a group of case studies. The first is on the complex diversity of farming systems in Tanzania, and the second, third and fourth report different aspects of work in the Amazonian floodplain of Brazil and Peru. There then follow a short report on diversity of upland rice and vegetables in a Chinese village, and an evaluation of an agroforestry practice in the same village. After this comes a report on the equivocal results obtained by use of household-by-household biodiversity assessment in another Chinese village. Then follows a discussion of the rapid transformation that is taking place in upland villages of northern Thailand. The last two chapters describe the changing situations in Guinée and Ghana, West Africa, and outline PLEC activities in support of the people.

Finally, Part IV is a short editorial statement of what has emerged from PLEC work relevant to policy in agricultural development.

2 Agrodiversity and agrobiodiversity

HAROLD BROOKFIELD

Introduction

THIS CHAPTER HAS the principal purpose of clarifying the distinction between agrodiversity and agrobiodiversity already remarked on in Chapter 1. Agrobiodiversity (or agricultural biodiversity) is the better understood of the two terms, and has a large literature. Agrodiversity is a term of wider ambit, and this chapter opens with a discussion of its meaning.

Agrodiversity

The term agrodiversity entered the literature only in the 1990s, largely through PLEC. Writing to introduce agrodiversity, and to distinguish it from biodiversity, Brookfield and Padoch (1994: 9) explained it as:

> the many ways in which farmers use the natural diversity of the environment for production, including not only their choice of crops but also their management of land, water, and biota as a whole. There is a close relationship between agrodiversity and managed biodiversity. Because of the diversity of cropping and resource systems that exists, agrodiversity serves as a major means of conserving both structural and species biodiversity.

Independently, Almekinders et al. (1995) also coined the term, and wrote of agrodiversity in arable systems as resulting from the interaction between plant genetic resources, the abiotic and biotic environments, and management practices. They defined it as 'the variation resulting from the interaction between the factors that determine the agro-ecosystems' (p. 128). The few others who have used the term, and whose writings I have seen, have done so without precise definition (Netting and Stone, 1996; Conelly and Chaiken, 2000). My own most recent attempt, in the course of a book written to explore the range and dynamism of diverse farming practices, draws on elements of previous definitions (Brookfield, 2001: 45–6). It runs:

> [Agrodiversity is] the dynamic variation in cropping systems, output and management practices that occurs within and between agroecosystems. It arises from bio-physical differences, and from the many and changing ways in which farmers manage diverse genetic resources and natural variability, and organize their management in dynamic social and economic contexts.

In the same book, I developed a classification first proposed by Brookfield and Stocking (1999). We identified four interrelated and overlapping elements of agrodiversity. The first was agricultural biodiversity, or *agrobiodiversity*, and the second was farmers' management of their soil, water, biota and other physical resources, or *management diversity*. Behind these lay the natural diversity of the physical environment, given stress by Almekinders et al. (1995), which we called *biophysical diversity*. Fourth, and by no means least, was the manner in which farmers and communities organize the use and allocation of their resources, and also their workforce. Often, this is marginalized as the 'socio-economic aspects' of farming systems. Agrodiversity gives it a much more central place, and we termed it *organizational diversity*. In this chapter, I deal only with agrobiodiversity. Chapter 3 takes up the questions of management and organizational diversity. In Chapter 4, Michael Stocking discusses biophysical diversity in more detail.

Agrodiversity quickly became a central concept for PLEC, because it was able to bring together the varied viewpoints on farming practices of the project's multi-disciplinary membership. Soil scientists could focus on the diversity of soils, botanists on that of plants, agricultural scientists on the range of cropping systems, and social scientists on the wide differences in social arrangements surrounding farming in the developing countries. Because all these aspects are interrelated, project members could work and talk together around the unifying concept. It aided recognition of the fact that diverse management of resources, diversity of crops, and management of trees, are systems developed by the farmers themselves.

We regarded 'agrodiversity as fundamental to an understanding of the sustainable adaptations by small farmers to varied environments ... [and] as fundamental to biophysical sustainability, principally through the high degree of structural, spatial and trophic, as well as species diversity that is involved' (Brookfield, 1995: 389). Project members found what they were doing to be both interesting and practically worthwhile. Meanwhile, another and larger movement was growing up, centrally concerned with only one part of this total diversity.

Agrobiodiversity

During the years in which PLEC was first being developed and during its operation, interest in agricultural biodiversity has surged. Its origins lay in the study of crop-plant domestication initiated in its modern form by Vavilov (1926). Domestication and the origins of agriculture have since garnered an enormous literature, summed up for the Old World by Zohary and Hopf (1993) and, more generally, by Harris and Hillman (1989). During the intervening years of research and germplasm collection of crops and their wild relatives, it became apparent that the replacement of locally evolved varieties (or landraces) by modern scientifically bred strains was leading to serious losses of landrace germplasm.[1] While the losses were greatest in those countries with industrial agriculture, they were perceived as being more serious in the tropics and subtropics, which have most of the world's plant genetic diversity. Juma (1989) calculated that 66 per cent of the genetic resources of food crops were derived from west-central Asia and Latin America alone. It is in the source areas, in the Near East especially, that the diversity of landraces and their weedy relatives was seen to be most severely imperilled. In the early 1970s, after two decades of rising alarm, major attempts were initiated to 'rescue' the endangered varieties of seed by collection and long-term storage. Within the framework of the network of International Agricultural Research Centres (IARCs), constituting the core of the Consultative Group on International Agricultural Research (CGIAR), a new body was set up in 1974 to coordinate germplasm storage internationally, including national storage and in the IARC facilities. Collection of germplasm extended throughout most countries of the world. By the mid-1990s over 6 000 000 accessions were stored *ex situ* worldwide, including wild relatives of the main crop plants in proportions ranging from 10 to 60 per cent. Storage conditions vary in quality and there is concern over the viability of many collections (FAO, 1996). Most of the securely preserved stock of seeds is held either in the IARCs, or in large well-equipped stores in countries of the North.

In the past two decades an alternative or complementary approach to protection of breeding material has been advocated. This is *in-situ* conservation on farmed land, mainly in the developing countries, where most farmers still replant seed they have saved themselves or have acquired from nearby. Comparatively few farmers still do this in the North. No country has a record of what is managed in this way, but there is certainly greater diversity, especially intra-specific diversity, managed on farms than is held in the genebanks. *Ex-situ* conservation is expensive, is dangerously subject to risk such as electric-power failures, and the samples are small, so that seed is not readily accessible to farmers (Cromwell, 1999). Critics of preservation in genebanks have also maintained that *ex-situ* conservation destroys the evolutionary dynamism that still continues on farms (FAO, 1996, 1998; Maxted et al., 1997; Jarvis and Hodgkin, 1998; Brush, 1999). Only in the 1980s was it fully appreciated how large an amount of crop plant diversity is still being managed by farmers, and especially by small farmers in the developing countries. Moreover, it was realized that whereas seeds can safely be stored in genebanks for long periods, the knowledge about how best to use them resides only with the farmers and their families. If varieties are lost from farms, this knowledge quickly disappears.

This was the background of the modern study of agrobiodiversity, the development of which has been closely linked with that of *in-situ* conservation. An important stage was the three-part 'Keystone' dialogue on plant genetic resources during 1988–91, which laid stress on '*in situ* conservation and on-farm management' (Keystone Centre, 1991). This had input into the work of the Convention on Biological Diversity (CBD) and into the FAO Global Plan of Action (FAO, 1996). By the mid-1990s, two international programmes concerned with the *in-situ* conservation of crop varieties on-farm had been generated. One was managed by the CGIAR International Plant Genetic Resources Institute (IPGRI) and the other by a consortium of non-governmental organizations (Manicad, 1996; Visser, 1998; Jarvis et al., 2000; Hodgkin and Jarvis, 2001).[2] From the mid-1990s, conservation of agricultural biodiversity on farmers' land has attracted a great deal of research and efforts to develop a practical set of methodologies (Jarvis and Hodgkin, 1998).

The focus of work in agrobiodiversity has been overwhelmingly on the plants themselves, and management is sometimes interpreted mainly as farmers' selection and acquisition of the germplasm that they plant in their fields and domestic gardens. Selection and acquisition attracted a lot of research interest during the 1990s, leading to some important findings, especially in areas of high varietal diversity of crops, such as potatoes, maize and rice (e.g. Brush, 1992, 1995; Bellon, 1996; Zimmerer, 1996; Louette, 1999). This, however, is only one part of management.

Agrobiodiversity can be a very wide field if interest is broadened to include all its aspects. As defined by two of its proponents, agricultural biodiversity includes 'all crops and livestock and their wild relatives, and all interacting species of pollinators, symbionts, pests, parasites, predators and competitors' (Wood and Lenné, 1999: 1–2). In most definitions, the field of study excludes wild animals, and plants of food or non-food value outside the agroecosystem unless they are wild relatives of useful plants. There is some confusion as to how far spontaneous biodiversity within the agroecosystem is also within the domain of agricultural biodiversity when, except as weeds, it has little or no actual or potential impact on agricultural production. This presents no problems for agrodiversity.

In fact, neither the scope of agricultural biodiversity nor that of the agroecosystem can readily be bounded. The question is fundamentally one of scale. When the scale includes whole ecosystems with an agriculturally managed component or the landscape, areas beyond the crop fields, wild fauna and plants other than the wild relatives of the crops become inescapably part of the domain of interest. The flora and fauna of uncultivated patches within and around the agroecosystem may not only provide important harvestable resources, but may also perform important indirect services to agriculture. Some of these services may be abiotic in function, for example in soil and water conservation. Some may be indirectly biotic, such as those harbouring a store of nutrients that will be used only much later. Wild biota have historically been of major importance both to human livelihood and ecosystem maintenance and remain so in many managed systems (Brookfield, 2001). Even in areas seemingly outside the agroecosystem, these wild biota are often subject to a significant degree of management: Chapter 3, and also some case studies in Part III, will expand on this aspect. Recognizing these problems of definition, one FAO-sponsored study enlarged the scope of the field so far as to interpret agroecosystems as being any ecosystem with human management (Aarnink et al., 1999). By this comprehensive definition, they could embrace the greater part of the whole biosphere.

The field of agricultural biodiversity, by definition, includes management not only of plant biota but also of soil biota. After several years of gestation, a new multinational GEF-funded project on below-ground biodiversity has come into being. The project is executed by the Nairobi-based Tropical Soil Biology and Fertility Programme (TSBF), and examines how above-ground crop and soil management affect soil biota. Inclusion of soil biota should imply that management of the soil itself is of concern. This question has been dealt with in some depth in the Wood and Lenné collection, with the complex and even contradictory findings from some 20 years experience of decline in the use of deep tillage in American and European farming systems (Edwards et al., 1999). The broad topic was taken up in a more theoretical context by Seybold et al. (1999). Management of the soil is a subject of rising importance in discussions of agricultural biodiversity, by no means yet adequately developed, or even properly appreciated.

Interrelations, actual and potential

Agrobiodiversity is a subset of agrodiversity, and in all parts of the PLEC project there has been detailed botanical inventory at landscape and field levels. The PLEC guideline document on biodiversity data collection (Zarin et al., 1999, partly reproduced here as Chapter 6) specifically described the task as the 'assessment of plant species diversity in complex agricultural landscapes'. Only a small part of the resulting work is reported in this book, because the analysis has taken a long time to complete. The interests of many PLEC scientists do, however, penetrate the field of agricultural biodiversity, and of conservation *in situ* on-farm. This becomes most specific in work done in China (Chapters 7, 18 and 20) and West Africa (Kranjac-Berisavljevic and Gandaa, 2000; Anane-Sakyi and Dittoh, 2001).

Much more work of this kind has been done outside PLEC. Moreover, a large volume of scientific research and literature has accompanied it. Three important new books on agrobiodiversity appeared in 1999 alone (Brush, 1999; Collins and Qualset, 1999; Wood and Lenné, 1999). By contrast, agrodiversity has until now produced far less literature. While diversity in farming practices is treated in parts of two important books of more general intent (Altieri, 1995; Gliessman, 1998), the first book specifically on agrodiversity appeared only in 2001 (Brookfield, 2001).

With its emphasis on plants, the literature on agricultural biodiversity has very little to say about the manner in which management practices vary between areas, communities or individuals. Agrodiversity, by contrast, gives much greater emphasis to the view at landscape level, and lays stress on management practices and their diversity among individuals. In putting emphasis on management, PLEC has addressed a gap left vacant by most of the modern literature on agrobiodiversity.

Conservation of landraces in the on-farm context raises big problems, because farmers' methods and crop choices can change rapidly in response to market signals and new introductions. All agroecosystems are subject to frequent or even constant change, and the principal interests of the farmers lie in the betterment of their lives, rather than in conservation. No agroecosystem is a museum either of crop plants or of field methods. To ensure farmers' cooperation in agrobiodiversity conservation, therefore, it has become necessary to seek ways of linking conservation with development (Almekinders and de Boef, 2000). PLEC has sought activities that would add value to biodiversity and encouraged these activities by material support. This has proved strikingly successful, and so have seed fairs and public occasions at which endangered species are displayed, and foods prepared from local resources are served. More detail on these incentive activities is presented in Chapters 10, 13, 23 and 24. Most farmers will conserve old landraces and old ways of doing things only if they can see advantages in so doing. No efforts to persuade them to serve voluntarily as museum conservators are ever likely to be successful for long.

It was only during the 1990s that farmers' efficiencies in site management, plant selection and even plant breeding came to be widely appreciated (Brookfield and Padoch, 1994; Almekinders and Louwaars, 1999; Brookfield, 2001). The intellectual development of both agrodiversity and agrobiodiversity has been strongly influenced by this recent realization of farmers' own abilities, potentially harnessed in association with scientific method, for improvement of crops and farming, and the conservation of biodiversity.

In the immediate future, it is probable that the specialized field of agricultural biodiversity will enjoy greater scientific and practical attention than the wider and more embracing field of agrodiversity. It is a subject that many of our colleagues find easier to handle, and it has a resonance to those who worry about the global impacts of diminishing biological diversity. Yet agrobiodiversity is incomplete without consideration of the management of the whole agroecosystem, and the lives of the land-users who are its guardians. It provides only a partial scientific basis, not the whole solution. In this book, we argue that attention to agrobiodiversity without considering its contextual position within agrodiversity not only misses good opportunities for protecting the global values of biodiversity, but also undervalues the smallholder farmers of the tropics and their management of a vital resource. Hopefully, a new structure for work on diversity will emerge in the coming decade. It should include the wider management and organizational aspects that are discussed in the following chapter.

3 Agrodiversity at the scales of farm and landscape

HAROLD BROOKFIELD

Introduction

AGRODIVERSITY ACQUIRES ITS full meaning and value only when the spatial scales of the farm and landscape are taken into account. It is at these levels that it most clearly provides context for modern agrobiodiversity, and contributes best to analysis of biodiversity conservation as a whole. The purpose of this chapter is to introduce the importance of these two scales for analysis; all the case study material in Part III relates to farms, agroecosystems, and the landscape. Discussion concentrates on management and organizational diversity, treating the field management of the biota as an integral part of farm management. Material from within PLEC provides most of the evidence, and this is supplemented where appropriate from outside the project. Methods of data collection for analysis are set out in Chapters 5, 6 and 7.

Diversity in management

Management in the arable fields

Anticipating the concept of agrodiversity in writing of central American small farmers, Wilken (1987) described their practices successively under the headings of the management of soil fertility and surface, of slopes, of water on and within the ground, of crop microclimate, and of horizontal and vertical space. All these are aspects of agro-technical management within agrodiversity, creating the conditions within which a diversity of crops is grown. PLEC has developed the concept of management further than Wilken. To provide a basic sampling structure for management diversity and biodiversity assessment, we distinguished two levels of analysis within landscapes in which a mix of local ecosystems and land uses is repeated in similar form over a wide area (Forman, 1995: 13). Chapters 5 and 6 set out methods for characterizing them in the field. Here they are summarized and placed in context.

First and larger is the 'Land-use Stage' or type, readily observable by its common land-cover characteristics. Second, and within the Land-use Stages, are the usually much smaller 'Field Types', determined by farmers' actual practices. Taken as aggregates, areas of arable fields, pastures, agroforests, planted or managed woodlands, fallows of different stages, and natural forest, would each be a Land-use Stage, recognizable from a distance, on an air photograph or on a satellite image. Field Types can be identified and classified only on the ground. They are of principal importance for the understanding of agrodiversity. Several of the chapters below, particularly Chapters 14 and 21, illustrate the detailed understanding of Field Types within Land-use Stages that can be derived using PLEC methodology.

The fine detail of site, crop and soil-fertility management on the slopes of Mt Meru in Tanzania is minutely described in Chapter 14. Up to nine distinct Field Types were identified within the one agroforest Land-use Stage, with all nine being present on one landholding. Moreover, the crop composition and layout in some fields is changed each season, so that two distinct Field Types may occupy the same piece of ground during a single year. Planting times are varied to optimize moisture and minimize pest attacks; irrigation and chemical supplements are selectively used; drainage needs are combined with soil-conservation needs; and manure and compost are applied in a timely manner. Those farmers with sufficient resources are able to deploy a wide range of strategies for soil conservation, soil-moisture improvement, and soil-fauna enhancement. The whole management system is not dissimilar to what is sometimes described as 'precision farming' in well-managed alternative agriculture in countries of the North (Committee on the Role of Alternative Farming Methods in Modern Production Agriculture, 1989).[1]

Methods of management in the field may be biological, physical or chemical. Work on the soil includes tillage, deep or shallow, manuring, fertilizing, irrigation and drainage by a wide range of methods. All these practices are designed to modify the surface and soil in favour of the crops to be grown. Biological management includes rotating or interplanting crops, weeding, encouragement of soil-improving plants and the preservation of other plants of value in the field or around its edges. Chemical management increased greatly in many areas during the life of 'green revolution' subsidies in the period between 1965 and 1990, but few small farmers can nowadays afford great expenditure on fertilizers and pesticides. Their use is generally quite limited and selective in terms of the crops to which they are applied. However, where a cash crop is followed by a self-provisioning crop in rotation, and the former has received industrial fertilizer, there is often a residual effect beneficial to the latter. Chapter 22 offers a specific illustration.

Farmers' practices include different levels of adaptive management of those biophysical conditions that they cannot change, and a more thorough adoption of innovations which seek to impose greater control over the farming environment. Enduring products of management innovations can extend even to changes in the physical or morphological conditions of the site. These include terracing the landscape to create new slope forms, and diverting the flow of water. They can extend to 'manufacturing' soils of a new type. Such major changes are sometimes termed 'landesque capital', defined as enduring improvements with a life beyond the current crop or crop cycle (Blaikie and Brookfield, 1987). Lesser changes, which can also be enduring, involve both harmful and beneficial variations in the ecological qualities of the soil. An important role of soil management is to ensure that physical qualities do not degrade or, if already degraded, can be restored.

Management among the trees

Field management also includes the deliberate interplanting of crops among trees, sometimes spared in clearing, sometimes springing up from the seed and root bank in the soil, or sometimes specifically planted for the benefits they provide. These can either benefit the soil by fixing nitrogen or, more generally, by nutrient-scavenging, or by their products, such as wood or fruit. The relation of trees to crops is a complex issue. In Ghana, as in many other regions, farmers distinguish trees that are favourable to crops grown close to them from others that inhibit the development of crops. The Ghana project group has initiated a programme of experimental investigation into the science behind these beliefs (Owusu-Bennoah and Enu-Kwesi, 2000). Farmers in the Embu region of Kenya have made similar observations to the PLEC scientists in that country.

Trees among crops vary in density from being scattered individuals to being so close and high that only the most shade-tolerant of crops can be grown. An example is reviewed critically in Chapter 19. In between is a large group of combinations collectively termed 'agroforestry', discussed in several of the following chapters. Its essential meaning is the deliberate association of trees with crops, with an immense range of spacing. It is now realized that these systems include a large number of soil-improving plants, as well as economically valuable plants, and that this range is constantly being widened as farmers experiment with whatever germplasm comes their way. Rural development science has only quite recently begun to take a serious interest in these many farmer initiatives, but they have been reported by scientific observers since at least the 1930s. From my own work and observations in the Chimbu area of Papua New Guinea in 1958–60, I wrote of the practice of planting *Casuarina oligodon* trees in fields still in crop, so as to provide wood and soil improvement during the subsequent fallow: *C. oligodon* has nodules and can fix nitrogen (Brookfield and Brown, 1963). Conklin (1957) earlier noted the importance of tree management in Mindoro, in the Philippines.

Systematic study of tree management in relation to crops has been undertaken since the early 1980s. The first work was done in the tropical regions of the Americas, where the cultivation stage of land-rotational systems is not seen as the centre of all activity, but as being 'in the midst of forest processes that are slowed for a few years so that a crop can be taken' (Alcorn, 1989: 70). In Peruvian Amazonia, a pioneer group initiated a major study of fallow management in the territory of the Bora people in the 1980s (Denevan and Padoch, 1987). Management extends so far into the fallow period that Denevan and Treacy (1987) wrote that:

The life of a field is one of sequential utilization rather than simply planting–harvest–abandonment–fallow. Harvesting proceeds from grain-producing
annuals (rice and maize) to root crops and pineapples to fruit trees and
unplanted useful trees and vines.

Management can continue for 10 or 12 years before selective weeding and
slashing finally come to an end. The effect over years could be a substantial
modification in the natural regeneration itself, as useful trees that were preserved in the fallow, by keeping clear spaces in an open canopy above them
until they were established, became a part of the continuing ecosystem.

Agroforestry is an ancient and very widespread practice, but for almost
two decades this fact has been ignored in much of the literature. For some
time after the early 1980s it was taken exclusively to mean alley-cropping
between a narrow range of nitrogen-fixing trees, usually exotics. This followed
experiments initiated by the International Institute of Tropical Agriculture in
Nigeria, and alley-cropping was vigorously taken up by other bodies worldwide later in the 1980s. The approach has been widely advocated both in
Africa and Asia, almost as a panacaea, to farmers both on flat land and on
sloping land, in the latter case with the additional object of soil conservation.[2]
But, even from the beginning, farmers were often reluctant. They found the
method not only burdensome in its labour demands, but also restrictive of
yields because of competition between the tree rows and the crops (Carter,
1995).

In the 1990s, the International Centre for Research in Agroforestry
(ICRAF), which had become the leading agency in the advocacy of alley-
cropping as a means of stabilizing shifting cultivation, began to recognize
these problems and to look at other and older forms of agroforestry, especially where soil-improving species were involved. This led to the
re-evaluation of sequential – as opposed to simultaneous – planting of nutrient-scavenging or nitrogen-fixing trees during fallow periods before and after
the crop, on the same land (Sanchez, 1995). Being not unrelated to the long-
derided 'shifting cultivation', sequential or relay agroforestry followed
ancient practices of farmers. The major change in approach was disarmingly
explained by ICRAF's then Director-General (Sanchez, 1999: 5):

> Competition for light, water and nutrients between the improved fallow and
> the crop is minimized by relay intercropping or sequential agroforestry sys-
> tems. Sequential improved tree fallow systems are more robust than
> simultaneous agroforestry systems, such as alley cropping.

In southeast Asia, a great many innovations in fallow management have been
developed by farmers, some of them a long time ago, some only recently. Most
include deliberate planting in the fallow, and sometimes planting among the
growing crops. The planted trees have either or both of two purposes. They
may improve the biological efficiency of the fallow process; they may make

the fallow more productive by introducing perennial economic species (Cairns, 1997; Cairns and Garrity, 1999). There is not much direct study of fallow management in Africa, but its spontaneous evolution in an area of Ghana, as part of farmers' own rehabilitation of land that had become degraded, was described by Amanor (1994). One suspects that, in at least some areas where fallow management still remains to be described, the problem may be one of its invisibility to observers (see Chapter 9), rather than actual absence.

Management of the fallow and the forest are treated in several places in this book, because PLEC has from early days followed a broad definition of agroecosystems. Chapters 9, 15, 16 and 17 show this most clearly and also emphasize the important role of management in areas of the Amazonian floodplain, or *várzea*, of Brazilian and Peruvian Amazonia. Enrichment planting and protection of desired species continue from field through to fallow stage, yielding managed fallows that are high in biological diversity and rich in tree species of value in a changing market situation. Fallows that are enriched with valuable species become forests, and they continue to be managed for production in this Land-use Stage (see Chapter 17). Farming here is quite specifically cultivating biodiversity, elements of which provide the main modern source of income.

Farmers' knowledge and management

Just as with agrobiodiversity, local knowledge, constantly modified by new information, is the foundation of all management practices. This is now widely recognized, and the potential utilization of farmers' knowledge for development has an extensive literature. In the area of soil and water conservation, in sub-Saharan Africa, farmers' innovations are now actively being sought (Reij et al., 1996; UNDP Office to Combat Desertification and Drought, 2000). But there is a danger in the modern enthusiasm for farmers' knowledge. It is not necessarily congruent with scientific understanding, since its categories and contexts are quite different. Some years ago, moreover, James Fairhead (1993: 199) warned against conflating good practice with what has come to be known as 'Indigenous Technical Knowledge'. He wrote that this is as reductionist as the older practice of 'conflating bad farming practice with "Indigenous Technical Ignorance". Arguably [he went on], all that has changed is the optimism or pessimism through which farmers' practices are observed.' In Chapter 11 of this book, Amanor raises another question, that knowledge is unequally distributed and is not often best obtained by 'focus-group' discussions dominated by a village leadership, seldom inclusive of the women or the very poor, and usually omitting also the sick, any tenants and other outsiders. Good knowledge, wisely applied, is the foundation of success, but poor knowledge, or failure to use good knowledge, is the foundation of serious problems. This is true in all farming.

Diversity in organization

The centrality of organizational diversity

Whether or not it sets out to make money, a farm is a working enterprise with its distinctive set of relationships with parallel enterprises, and the higher levels of the community, the authorities, and the regional, national and global economies. Like any other enterprise, it is both a social and an economic system nested within larger social and economic systems. The operator of the farm is the land manager, in the sense used by Blaikie and Brookfield (1987). Even if he or she has to work within a system which determines what crops and livestock are produced, the farm operator has to take the yearly, monthly and daily round of decisions needed to obtain that production. Farms differ greatly from one another, and the resources and skills of farm operators also differ greatly.

This is a central part of diversity, and in PLEC we have used the term 'organizational diversity' to characterize it. In this chapter I give it particular stress. It includes diversity in the manner in which farms are owned or rented, and operated, and in the use made of resource endowments and the farm workforce. Elements include labour, household size, the differing resource endowments of households, and reliance on off-farm employment. Also included are age-group and gender relations in farm work, dependence on the farm as against external sources of support, the spatial distribution of the farm, and differentials between farmers in access to land. Tenure of resources, the conditions of access to them, and what Leach et al. (1999) describe as 'environmental entitlements' are fundamentally important. Organizational diversity is involved in all management of resources, including land, crops, labour, capital and other inputs.

Land tenure is a particularly important issue, not only concerned with the amount of land farmers hold, but also the conditions under which they operate their farms. In some areas, among which the Fouta Djallon of Guinée is a striking example, where a large part of the land is held by very few people and almost all the outfields are rented or share-cropped (see Chapter 23). In China, notwithstanding the redistribution of land in the early 1980s under the Household Responsibility System that succeeded the break-up of the communes, a significant amount of tenancy had re-emerged by the end of the 1990s (see Chapter 7). Tenancy is also very significant in southern Ghana, as explained in Chapter 24, but here an initial conclusion, that the land of tenants was more degraded (Gyasi, 1996a), was overturned by more detailed work (Gyasi, 1999). A number of tenants had started conserving trees on their land well ahead of the PLEC campaign for tree conservation, and managed their land with great care. Some had become leading members of a PLEC farmers' association. This observation is not unique to Ghana, and it is important because the scientific (and policy) view is that absolute security of tenure is essential in order to permit farmers to make investments in their land. In the currently dominant neo-liberal (free market) development paradigm, privatization of land ownership is an essential precondition for advance. This view disregards problems of rising social inequality, which are discussed below.

Whatever the conditions of tenure, the skills required in simple organization of the workforce at periods of peak demand are considerable, and are much undervalued in the general literature on agricultural development. The shift from single to double, or even triple, cropping made possible by the 'green revolution' innovations was enormously demanding of these skills. I once saw the relay stage between crops on a small rice farm in Java. Within not much more than one hectare were some still-standing crop, some harvested rice being threshed, a new seedbed sown broadcast that morning and being flooded in the afternoon, and another field being prepared for the new crop with a buffalo-drawn plough (Brookfield, 2001: 232). Yet although Javanese farmers in that period were subject to close direction, neither they nor any other developing-country farmers received much guidance and instruction on how to manage their resources and workforces at such times. They did this by themselves. Organizational diversity is highly dynamic. Farmers change their organization of labour and resources according to circumstances, sometimes in a very short space of time, and are quick to respond to signals which call for new ways of combining the factors of production.

Social differentiation in relation to management

An important element of diversity in the organizational sphere is social differentiation. Richer farmers are more able to invest in their land, and experience less risk in trying out innovations. Whether or not they actually control the labour of poorer farmers and the landless among them, they have the resources to hire labour when they need it. By renting out land on a cash or share-cropping basis, the landlord-farmers gain part of the produce of their tenants. Where there are actual patron–client relationships, the wealthier farmers may sometimes give support to their poorer neighbours in need, and in return may be able to use the land and labour of the poor to support their own enterprises. In yet other cases, and there are many of them in the world, there are established social mechanisms for the levelling of these differentials in access to production and incomes. Social networks are often of major importance.[3] Chapter 4 provides detail from a study of conservation management in Kenya that reflected the different resource endowments of farmers (Tengberg et al., 1998).

The relation of the maintenance of diversity to social differentiation is by no means clear and straightforward. In the Paucartambo valley of highland Peru, two social classes of farmers maintained high levels of landrace diversity among their potato and maize crops. The very poor did this because they had no access to the land on which commercial crops could be grown; the rich did so for reasons of dietary preference, for use as gifts, and for payment of workers (Zimmerer, 1996). In a high-population-density area of western Kenya, the Luhya people as a whole have adopted a very complex agrodiversity as the population has grown, and land on which to practise shifting cultivation and graze animals has grown scarce or vanished (Conelly and Chaiken, 2000). Principal elements in diversity have included intercropping among the staple maize and sorghum, cultivation of a wide range of landrace varieties, giving emphasis to crops that have multiple uses, especially in relation to livestock, and the cultivation of field crops among tree crops. Necessarily, livestock numbers have declined.

All farmers responded to the growing constraints by matching crops and landraces to particular soil types and locations, and maintaining recycling of manure and wastes on the land closest to the home. The weakness was that many farms were so small that they could not support those who lived on them for more than a small part of the year, and in the studied area all the people had protein-deficient diets, and had to buy a significant proportion of their food, making investments in production possible only for those with access to good incomes, mainly earned by family members working away from home. Conelly and Chaiken (2000: 39) conclude that 'despite the resourcefulness and diversity of their agricultural system, the overall picture . . . is of a people who have marginal diets and lack basic food security'.

Also among the Luhya, at a population density over 1000/km^2, only a short distance from where Conelly and Chaiken worked, Crowley and Carter (2000) found that most farms were unable to employ timely fallow, crop-residue incorporation, crop rotation, or purchase of soil fertility inputs. They were unable to devote adequate labour to what little land they held because they needed to devote so much of their time to earning the income necessary for survival. This pattern is relatively unusual, although not unique. In many areas, those who hold little land place much greater emphasis on diversity and intensive management than those who hold more adequate land. This is also the case in Yunnan, southwestern China, although the leader in diversification in one village is also one of the most prosperous farmers (Chapter 21). Diversity in the ownership of land and access to capital assets are of major importance in helping to explain management diversity and its variation between particular farms and communities.

Gender, expert and inexpert farmers

Gender differentiation is of great importance, since it is only rarely that field tasks are fully interchangeable between the sexes. In some areas with high male absentee rates, more than 90 per cent of farm work can be done by women. In these circumstances, gender division of labour is often highly flexible, but land tenure is not always so accommodating to social needs. Women both own and operate land in Brazil, but where women can own land in Africa, it is often in smaller parcels, and inheritance may be only in the male line. Among the Fulße of Guinée, women have secure tenure to their own home plots of land only while married (Chapter 23). In southern Ghana, where there is much share-cropping tenancy but only a minority of female farm operators, women operators are more constrained than men in their land- and tree-management decisions. Many opt out of active farming in favour of processing work, which is more traditionally a female area (Ardayfio-Schandorf and Awumbila, 2000). The more positive response to their constraints by one group of women farmers in central Ghana is described in Chapter 13.

The question of expert farmers and their identification is taken up mainly in Chapters 8 and 12. Since its early days, PLEC has worked increasingly through these selected individuals, most of whom are not the recognized community leaders or those who have most faithfully followed the advice of the extension services (Pinedo-Vasquez, 1996; Chapter 12, this volume). They are also not the ones who habitually participate in external projects. In all PLEC areas the identified expert farmers include women who have rarely participated in external projects. The converse of the identification of expert farmers is the unsurprising fact that a proportion of farmers in any population is likely to be less than expert in relation to their production goals, or the conservation of useful biodiversity. In this book a Chinese example of such a situation is presented in Chapter 20.

Change in the agroecosystem

Everything in agricultural systems changes through time. There is the cycle of seasons, and the great variability that almost always makes this cycle far from regular. There are secular climatic changes, and frequent events such as those associated with the El Niño and La Niña phenomena, that have major impact regionally. In all systems there is intra-seasonal and inter-seasonal variation in decision making on the use of land, labour, capital and other resources, and in the security or risk of the harvest. Longer-term changes in cropping and management practices occur as soils and biota are modified by use and natural processes, as self-provisioning gives way to commodity production, as markets open up or fade away, and as new crops are adopted and others discarded. Farmers constantly experiment with new introductions, and with new ways of doing things. The old view that 'indigenous' farming systems are static, and have endured with little change for a very long time, has increasingly been challenged by a view of farming systems as dynamic, whether rapidly or only slowly changing.

The relevance of the 'new ecology' to the interpretation of change

Agroecosystems are ecological systems, and their evolution through time involves processes that lie in the domain of plant ecology. It is often said, by no means always correctly, that agroecosystems are simpler than wild ecosystems, with less diversity, and usually a less complex structure. Until recently, they have therefore been thought of as more unstable than wild ecosystems, and more liable to degradation unless well managed. In the early days of modern ecology, Clements (1916) proposed a model of climax vegetation, and unilinear succession toward it after disturbance, that has had remarkable influence and persistence. In ecology, it was not seriously questioned until the 1960s. By that time it had been joined by the development of cybernetic ecosystem ecology which powerfully enhanced the notion that once a climax had been attained, ecological systems tended to resist change, maintaining their stable state homeostatically (Odum, 1954). If, on the other hand, disturbance was not relaxed, or was repeated after only a short interval, the consequence would be resource degradation. These views, like the views of 'traditional' farming systems as being in balance with the ecological environment unless or until disturbed, have now been overtaken by different ideas. But the long shadow of these old ideas still falls across a large part of the literature.

Ecologists now recognize frequent and widespread disturbance as integral to the structure and content of ecosystems, even as necessary to sustain biodiversity. Disturbance can be natural, or human induced, and it occurs on a huge range of scales both in time and space. In place of the old notions of homeostatic equilibrium, modern thinking is in terms of unstable but resilient systems capable of renewal, but with variations, after major disturbance. Although a succession always follows disturbance, it can lead in many directions, and the emphasis has increasingly been on the dynamics of patches within plant assemblages of all stages. Disturbances create patches, and disturbances of different types have different effects.

During the 1990s, new ideas concerning non-equilibrium dynamics have increasingly been applied to the study of change in agricultural systems (Scoones, 1994, 1999; Zimmerer and Young, 1998). Fields, fallow areas, edges and wild growth are all patches, each with their own dynamics, which are managed by farmers. Revising earlier ideas about trends in the African Sahel, Mortimore (1998) saw instability in the biophysical system being coped with by unstable, yet highly resilient, adaptation in what are here called management diversity, agrobiodiversity and organizational diversity.

The significance of resilience

Agrodiversity fits much more naturally into the concepts of the 'new ecology' than into those of the old. It eschews any notion of automatically regulated, equilibrium-seeking systems in any of its elements, natural or social. It gives central recognition to adaptability, farmers' experimentation and change. Its analysis requires a recursive exploration into dynamic processes in the biophysical system that is being managed, and the social and political context of management, as well as in the technical content of crops and methods.

An emphasis on disturbance is also an emphasis on uncertainty and complexity. Both in the managerial and organizational spheres it calls attention to the concept of resilience, already introduced above, and the associated concept of sensitivity. Varying the classic definition of resilience by Holling (1978), a soil, a vegetation complex or a social system are resilient if they can utilize and even benefit from change by quickly adopting a new productive or operating condition, whether or not it is similar to the pre-disturbance condition. Sensitivity is a measure of the amount of disturbance needed to change an existing condition, but a soil that does not resist disturbance, but quickly regains a productive condition after it, is resilient (Seybold, et al., 1999). Similarly, while it is a rare farming society that can shrug off changes in the external forces impacting its operation, many individuals in such societies are demonstrably able to make organizational changes that create an opportunity out of the disturbance. The seizing of opportunities that the time offers is not only the path to success; in the developing countries now, as was said in Japan in 1822, it may be the key to survival (Okura Nagatsune, cited in Smith, 1988: 212).

The analysis of agrodiversity is opposed to generalizations about unidirectional change based on snapshot observations and assumptions used without empirical evidence. It is not assumed that successes and failures in one area are typical of all, but it does emerge that dynamic and contrasted trends are to be expected. Moreover, generalizations about whole regions and their people are seldom, if ever, true of all farms and individuals within them. This emerges strongly in several of the case studies presented in this book. To explore this everywhere might be unproductive in a context in which the aim is to influence policy. 'Exposing complexity and difference can easily leave policy makers cold' (Batterbury and Bebbington, 1999: 285). But while it may be a tall order to elucidate the interplay of social and natural dynamics everywhere, it improves the quality of written output to go into the field expecting varied responses to be found. Increasingly, as its work has advanced, PLEC has learned this lesson.

Agrodiversity has demonstrated striking resilience through the centuries, and it still does (Brookfield, 2001).[4] Its resilience certainly relies on diversity in agrotechnical management, but much more on the adaptability and innovativeness of the farmers, and especially the expert farmers among them. It is quite wrong to regard small farmers as dangerously vulnerable to all forms of modern externally-forced change. This view, which is widespread, is based on an impression of small farmers as conservative and unwilling to innovate except when under pressure to do so. This book demonstrates that the reality is very different. Farmers are very adaptable and often innovative. They seize opportunities readily. It is the adaptive dynamism of agrodiversity that underlies its resilience, and is its most essential property for survival.

4 Agrodiversity, environmental protection and sustaining rural livelihoods: the global view[1]

MICHAEL STOCKING

Introduction

THE UNITED NATIONS Convention on Biological Diversity, signed at the Earth Summit in Rio de Janeiro in 1992, and ratified by most developed countries in 1994, set out a global agenda for the conservation and sustainable use of biodiversity and its components. It affirmed the sovereign rights of states over their biological resources and, crucially, it stated a principle that the benefits of development of biological resources should be shared fairly and equitably on mutually agreed terms. This chapter is about these benefits of biological diversity. It concentrates on the benefits for disadvantaged people – those living in degraded, poor and marginal environments, for whom global benefits may seem an irrelevant luxury. It highlights the situations of those living in underdeveloped, often tropical, parts of the world, for whom development aid is usually targeted.

Biological diversity has traditionally been a central concern of ecologists. Yet for many non-ecologists, it remains conceptually elusive (Rodda, 1993). Treated with suspicion by some, biodiversity means different things to different people. Some observers look upon it as a bandwagon for their pet agendas, while others bemoan the loss of precision and rigour in debates on conservation priorities (Blaikie and Jeanrenaud, 1997). Referring to the variety, or the number, frequency and variability of living organisms and the ecosystems in which they occur, biodiversity itself encompasses diverse aspects. It includes:

○ species diversity, which is the number and frequency of species of plants, animals and micro-organisms
○ intra-species and genetic diversity, the variety between individuals of the same species and of genes within each species
○ ecosystem diversity, the different types of animal–plant assemblages and their associated ecological processes.

For many years, assertions as to the threats to humans and global life-support mechanisms by loss in biodiversity (e.g. Wilson and Peters, 1988), and supporting arguments on the benefits and potential values of biodiversity (e.g. McNeely, 1988), always came from ecologists. It was only in the early 1990s, with the rise of ecological economics and political ecology (e.g. Swanson,

1995), that biodiversity entered a very different arena – one that would attract pragmatic, immediate and purposeful reasons why protection of biodiversity is essential. The 1998 Third Conference of Parties to the Convention on Biological Diversity (COP3/CBD), meeting in Bratislava, asked countries to document and demonstrate the importance of biological diversity in the course of addressing problems of resource depletion and of supporting rural communities. This chapter arises from this fundamental shift in thinking – from biodiversity being an ethical but esoteric concern of natural science, to biodiversity being a foundation for human livelihoods and sustainable development.

If we are to take up the challenge of COP3, we need to demonstrate the central importance of biological diversity in addressing both the old problem of the depletion of environmental resources and the more newly articulated problem of rural people's livelihood and welfare. PLEC does this through the present book. This chapter presents the global view. It brings together three terms – agrodiversity, land degradation and rural livelihoods – that are all receiving substantial, but usually separate, attention in current debates about global futures and how human society can attain sustainable development.

A framework for sustainable rural livelihoods and environmental protection

In parallel to the change in thinking about the importance of biodiversity, approaches to human welfare in relation to the biophysical environment have also undergone radical shifts (World Bank, 1992). No longer is development simply the transfer of technology or the delivery of economic support from an altruistic donor to a grateful recipient. It is far more complicated. Yet there are still searches for some 'magic formula' that will ensure that the agendas of both the recipients of aid (e.g. economic development and poverty relief) and the givers (e.g. biodiversity conservation) are met. One such is the appealing notion that the biophysical environment can be looked after while simultaneously providing for the immediate and felt needs of human populations – the World Bank's 'win-win' scenario. It has driven international development writing on some of the seemingly most intractable human problems, such as population growth and environmental decline in Africa (Cleaver and Schreiber, 1994). In part, the notion is a reaction to the widespread environmental pessimism of the 1960s and 1970s. Partly, also, it has been fuelled by evidence that environmental recovery has, in places, happened despite population growth, and even because of it (Tiffen et al., 1994; Critchley, 1999).

To cite one small example, PLEC work has shown that some traditional soil conservation methods, such as trashlines in drier parts of Kenya, maintained against professional advice that would rather have terraces dug, can be both economically and technically efficient (Kiome and Stocking, 1995). If such positive things have happened without (or even despite) planning, then imagine – the argument goes – what could happen if the ingredients for a 'win-win' scenario could be identified. Their effects could then be analysed and the techniques could be appropriately supported. Much attention has been given in the literature to the search for alternatives, asking questions such as: 'Can agricultural growth be compatible with natural resource conservation and at the same time create the conditions for improving livelihoods and reducing poverty?' (Ellis-Jones, 1999: 179). Much of this attention has been devoted to indigenous technologies, especially in the more difficult, arid parts of the tropics (Kerr and Sanghi, 1992).

Figure 4.1 presents a 'win-win' scenario for sustainable rural livelihoods, with components involved in land degradation and biological diversity. Environmental protection is essentially part of the global agenda, enshrined in two global conventions and reflecting primarily the needs of society. Many land users also wish to protect their environment but this is in the context of their primary goal of maintaining production rather than for environmental protection *per se*. PLEC recognized this from the outset, and almost all the case studies in Part III of this book demonstrate ways in which the enhance-

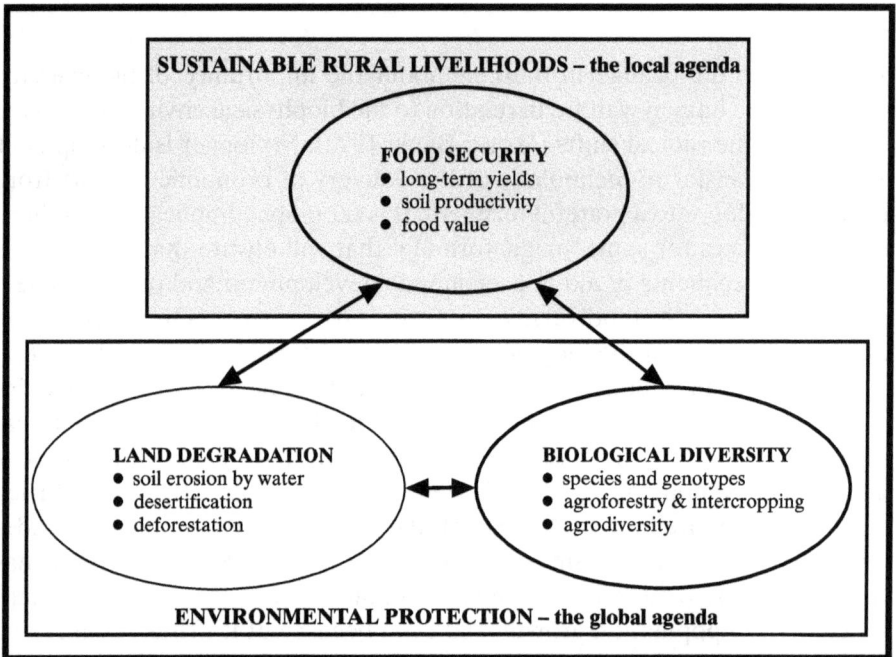

Figure 4.1   *A 'win-win' scenario. Situating agrodiversity and land degradation in the sustainable rural livelihood debate and global environmental agendas*

ment of biodiversity can be combined with improvements in production and livelihood. The rest of this chapter will look first at these two elements of the global agenda and how attention to one may benefit the other. It will use evidence from PLEC and other recent research to show how these two global agenda items may address sustainable rural livelihoods (SRL) for rural people.

Land degradation – part of the global agenda

Under the UN Convention to Combat Desertification (CCD), desertification is defined as land degradation in arid, semi-arid and dry sub-humid areas.[2] Article 1 of the CCD describes 'land' as the terrestrial bio-productive system that comprises soil, vegetation, other biota, and the ecological and hydrologic- al processes that operate within the system (World Bank, 1996). 'Land degradation' therefore means the reduction or loss of biological and economic productivity of major land uses (rainfed cropland, irrigated cropland, range- land, forest) resulting from processes arising from human activity, such as soil erosion, changes in physical, chemical or biological soil properties, and long- term loss of natural vegetation. In giving priority to Africa, the CCD echoes the common belief that the most intractable degradation problems are in the dryland agricultural and livestock practices of Africa's subsistence land users.

Land degradation causes problems at three levels (Pagiola, 1999). At the field level, it results in reduced productivity. At the national level, land degradation has large off-site effects, including flooding, sedimentation, effects on water quality and reductions in water supply. In developed coun- tries, such as the USA, these off-site impacts are calculated as being many times larger than on-site changes in soil productivity (Crosson and Stout, 1983). In less developed areas, such as Java, the opposite is believed to per- tain (Magrath and Arens, 1989). Caution does, however, need to be exercised in these national estimates, because an economic valuation approach and the perspective adopted substantially affect such calculations (Enters, 1998).

At the global level, land degradation has a number of complex direct and indirect interaction effects, including climate change through emissions of greenhouse gases, reduction in terrestrial carbon sinks, diminished effective- ness of carbon cycling, reduction in biodiversity, and pollution and sedimentation of international waters. Many of these global aspects of land degradation have major implications for its control through the use of bio- logical means. This link is explored first in the context of biodiversity, and second in the context of sustainable rural livelihoods.

Biological diversity – the global agenda

Biodiversity is characterized by variety and variability. Although usually asso- ciated with wild nature and protected areas, the CBD has encouraged a much broader and pragmatic interpretation of the term to mean the variety of all life on earth, whether wild or domesticated (Koziell, 2000). This variety is both spatial and temporal. Living things interact among themselves, and also with the inorganic environment of atmosphere, soil and water. Human society taps into these interactive relationships, primarily through its utilization and modification of food chains, and the attendant processes such as photosynthesis. Biodiversity is fundamental to life-support, sustaining the interconnectedness of life and adaptation to changing conditions. Table 4.1 presents a partial view of how different components of biodiversity are significant to life-support, showing how the human significance may impact at all levels from the local to the global.

Table 4.1   Types of biodiversity and their human significance

Type of diversity	Human significance	Individual components
Ecosystem diversity – biodiversity at landscape level, with distinct groupings of living things.	Interactive processes within and between species, providing ecological resilience; interactions with physical environments, enabling hydrological and nutrient cycling, and major accumulations of energy.	Species combinations at multiple levels; from ecosystems occupying different parts of landscapes, to species within a single ecosystem.
Species diversity – biodiversity in its classical sense, being the distribution and abundance of species within a community or an area. Includes species richness.	Provision of a range of biological resources, required for survival. Includes food, fodder, and fuel. Reduction in risk if one species should fail.	Variety in species of birds, mammals, invertebrates, plants, fungi, micro-organisms. Subspecies and varieties, including cultivars, landraces and domesticates.
Genetic diversity – diversity in frequency and variety of genes and/or genomes within populations and species. Includes differences between genes on the same chromosome.	Creates hereditary variation within species, offering opportunity to choose genetic characteristics which relate to yield, taste, palatability, nutrition, drought resistance, etc.	Alleles, phenotypes, DNA sequences.

Adapted from Koziell (2000)

The presentation in Table 4.1 of the importance of biodiversity to human society is a somewhat standard view. Many observers, even those working directly on biodiversity issues, are sometimes dismissive of the conservation of biodiversity as 'important only up to a point for the sustainability of agricultural production' (Tisdell, 1999: 9). Different authors variously advance five main reasons why biodiversity is so important to human society:

o *greater functionality*: diversity within and between species gives them differing characteristics, enabling more roles and functions to be undertaken

○ *inter-generational equity*: future human societies may become dependent on living resources that are preserved now, without apparent current productive benefit
○ *better resilience and stability*: agroecological systems are more resilient and sustainable if they possess greater genetic diversity, though empirical evidence is far from conclusive (Perrings et al., 1995; see also Chapter 3.)
○ *flexible response to changing needs*: biodiversity enables species to adapt to changing conditions and external pressures
○ *ecosystem and life-support service role*: energy conversion and material exchanges are necessary for maintaining a productive, healthy and stable environment.

Biodiversity, then, is the source of many goods and services for society at a variety of functional levels. Of particular practical interest at the global and international levels are:

○ conservation of genetic material of future importance to society, including medicines and chemicals not yet discovered
○ preservation of genetic material currently useful to society but not yet widely disseminated, or under threat
○ risk reduction in material provision during times of environmental stress and scarcity
○ production benefits through selection of more productive species and varieties
○ diversification by increasing global production of a greater range of crops.

Returning to the 'win-win' scenario in Figure 4.1, biological diversity should be well equipped to provide benefits both to society at large and to local people. Most of the benefits considered above can equally be considered as positive aspects for rural people. Furthermore, they are aspects that are maintained by them, when they manage their biophysical resources of plants, soil, water, climate and land. This is what PLEC describes as agrodiversity. But before considering these complex linkages, the final component of the 'win-win' framework is sustainable rural livelihoods (SRL).

Sustainable rural livelihoods (SRL) – the local agenda

In the past five years a new focus for development aid assistance has arisen – the SRL approach. It is being embraced by development agencies such as the UK's Department for International Development (Carney, 1998). The approach is distinct from previous donor efforts such as 'Integrated Rural Development Planning' of the 1970s and sectoral project planning of the 1980s. Above all, it is people-centred, rather than focusing on geographical areas or needs and resources. It targets poverty, complexity, local knowledge and skills, while attempting to relate policy to practice. It takes the opportunity to mainstream environmental concerns (such as land degradation or loss in biodiversity) as part of livelihood development, rather than just adding them as further donor-led requirements. A framework for SRL has recently been developed in order to concentrate the focus on people, their needs and their opportunities (Scoones, 1998). Intended to be a simple but practical tool, the framework concentrates on five different assets upon which individuals may draw in order to build their livelihoods (Table 4.2). The SRL framework gives a useful means of organizing the many types of information relating to the land user, the production system and their potential influence on land degradation. In particular, the framework can highlight circumstances that make land degradation a possible outcome of future activities, or where the transfer of 'capital' from one type to another may affect the potential for degradation, or the possibility of environmental protection. Such scenarios have been constructed for a number of detailed examples (Stocking and Murnaghan, 2001), in order to show how the capital assets information may be used in practical field situations.

Table 4.2 develops examples of how the availability or unavailability of capital assets may affect land degradation positively or negatively. Thus, for example, one important asset component affecting land degradation is 'social group'. Disruptions to society through conflicts and wars are well known to cause environmental stress through refugee movements and people's unwillingness to invest in land improvements when the future is so uncertain (Black and Sessay, 1997). Alternatively, long-settled and cohesive communities and social groups tend to develop their own tried and tested technologies, including a wealth of conservation measures (Tengberg et al., 1998). Table 4.3 sets out the capital assets components of SRL.

Another example of the new approach is to be found in the *gestions de terroirs* of West Africa. Loosely translated as 'management of village lands', projects using the *gestions de terroirs* approach start with the participatory identification of resource problems. Guèye (1995) describes the types of tools used, such as resource maps, Venn diagrams and wealth classification, as well as the use of local knowledge. This leads to the development of village institutions to address resource management decisions and implementation, monitoring and evaluation of activities (Toulmin, 1993). The approach emphasizes a number of important aspects of sustainable rural livelihoods:

○ local diagnosis of problems – using those who live with the problem to articulate a need to address the problem
○ establishment of boundaries of responsibility – those who do the work reap the benefit

Table 4.2 Examples of capital assets components related to land degradation

Capital asset	Component	Positive relation to land degradation	Negative relation to land degradation
Natural	Land	Sufficient lands to plant grass strips or shelter belts	Insufficient land to devote to non-consumptive uses such as conservation
	Soil	Resilient soil able to withstand continuous cultivation without yield loss (e.g. Nitosol)	Sensitive soil that loses fertility easily, which has large impact on yields (e.g. Luvisol)
Social	Social groups	Indigenous and adapted techniques of soil and water conservation developed	Social disruption, conflict, wars
	Access to institutions	Effective agricultural extension and subsidy regimes	Ineffective agricultural extension and lack of access to technologies
Human	Labour	Sufficient labour to intensify and invest in soil improvement	Extensification and poor agricultural practice because of insufficient labour
	Gender	Women's management of home gardens using composting, ridges and intercropping	Overburdened women unable to undertake soil improvement
Physical	Implements	Availability of ridging implements to construct earth bunds	Conventional tillage only, encouraging excessive erosion
	Transport to market	Market opportunities provide further investment in land	No cash income – subsistence cultivation only and soil mining
Financial	Cash income	Sufficient cash to invest in land and soil improvement	Non-farm cash income sources discourage investment in soil
	Credit	Land users enabled to plant on time and minimize erosion	Intermittent crop failure, debt and necessity to seek off-farm income

○ development of a village management plan and map – to encourage a continuing involvement and recognition that needs in one place differ from those in another

○ regular monitoring and evaluation – an iterative cycle of learning and adjustment of activities.

Gestions de terroirs stand a far better chance than earlier methods of accurate diagnosis of land degradation problems and committed land management. They are also more likely to rely on the intrinsic biological resources available at village level, rather than external interventions and

Table 4.3 Capital assets components of sustainable rural livelihoods (SRL)

Capital assets	Attributes
Natural	Natural resource stocks from which resources flow to support livelihoods
Social	Social resources from which information, mutual help, emotional support and physical assistance flow
Human	All directly productive aspects deriving from humans, such as intellectual skills, knowledge, physical strength. Labour and gender are two that feature most in determining investments in the land
Physical	The basic infrastructure, and the production equipment and means, such as tools, transport, energy and access to water
Financial	The financial resources available to people, such as credit, savings, pensions and cash income

After Carney (1998)

imported inputs. While not regarding *gestions de terroirs* as a panacea, Reij et al. (1996) consider it a more promising way forward than standard solutions. Village-based land management highlights the need to combine all material, social and physical resources towards achieving sustainable support to livelihoods. It thus shares many of the attributes of the PLEC approach, as previously discussed in Chapters 1, 2 and 3. The difference is that PLEC does not emphasize 'participatory rural appraisal' and its related techniques as precursors to action. These are part of the formal science paradigm where reductionist understanding is needed before learning can begin. PLEC starts with its scientists learning from local people.

Agrodiversity

Biological diversity has a crucial ecological role in agroecosystems (Altieri, 1999). In addition to food production, it performs ecological services in nutrient cycling, regulation of microclimate, suppression of undesirable organisms and detoxification of noxious chemicals. Most authors (e.g. Thrupp, 1998) see this as a part of 'agricultural biodiversity' or 'agrobiodiversity', as discussed in Chapter 2 . PLEC calls it 'agrodiversity' and many of its attributes and values are discussed in Chapters 2 and 3. Agrodiverse farming systems are resilient to external pressures and are able to provide assured and sustainable production through the local management of biological diversity.

There are several pragmatic reasons why agrodiversity should be a useful focus of interest today. First, there is substantial and economically important plant biological diversity in areas of land use in the tropics and sub-tropics. Areas of land use encompass most of the places where natural biodiversity is at a maximum, while protected areas tend to occupy more marginal zones. As human populations increase, protected areas will become an increasing luxury and hence even their role in maintaining the stock of biodiversity will be under threat. Therefore, it makes sense to concentrate efforts on keeping biodiversity in agricultural systems. Second, small-farm agricultural systems are the primary custodians of this biological diversity. Project surveys have consistently shown the wealth of this knowledge. For example, in one region in southwestern China, Guo and Padoch (1995) revealed 82 forms of managed agroforestry and 220 different associations of trees and crops.

In such situations, local adaptation has effectively maintained biodiversity and even enhanced it. Unfortunately, much of this knowledge and management of biodiversity is slipping away under the pressures to commercialize, mechanize and standardize. In particular, only the elderly now hold knowledge of local plants such as medicinal species; chemicals have superseded many locally developed pest management practices; and many soil fertility maintenance techniques are now reliant on inorganic inputs. There is an increasing recognition that agrodiversity might hold lessons for agricultural development, especially in accessing techniques that would improve the situation of the rural poor.

Agrodiversity is largely a response of resource-poor farmers to inherent spatial and temporal environmental variability (e.g. the case of farmers in semi-arid Kenya; Tengberg et al., 1998). Brookfield and Stocking (1999) have codified agrodiversity into four principal elements, all of which have major implications for livelihoods and environmental protection: these categories are discussed in Chapter 2. Here, they are placed in a different order, and slightly reinterpreted, in order to suit the present argument better.

○ *Biophysical diversity* of the natural environment controls the intrinsic quality of the natural resource base for food production. It crucially involves knowledge of earth surface processes and interactions between organic and inorganic components, for the control of land degradation by water erosion through vegetation cover.
○ *Management diversity* involves all methods of managing land, water and biota for crop production and the maintenance of soil fertility and structure.
○ *Agrobiodiversity*: this is interpreted in this chapter as the management and direct use of biological species, including all crops, semi-domesticates and wild species. In agricultural systems, the diversity of crop combinations, and the manner in which these are used to sustain or increase production, reduce risk, and enhance conservation, are aspects of agrobiodiversity important to livelihoods and environmental protection.
○ *Organizational diversity*: the way in which farms are owned and operated. Farmers with different resource endowments organize differently in accordance with their specific circumstances. Flexible responses to external events and market opportunities facilitate better support for livelihoods.

Brookfield and Stocking (1999) argued that these categories are fundamental to understanding the interface between natural biological diversity and human land use. This management of variation within agroecosystems (Almekinders et al., 1995) is increasingly acknowledged as the principal way forward in order to achieve the 'win-win' scenarios such as that described in Figure 4.1. Two benefits of agrodiversity will now be considered – control of land degradation and promotion of food security.

How agrodiversity controls land degradation

Through the greater use of biological processes and of techniques adapted by farmers in response to the local environment, agrodiversity reduces both the risk of land degradation and its potential impact on production and livelihoods. This is achieved in complex direct and indirect ways through combinations of:

○ better and more continuous vegetation cover intercepting rainfall erosivity
○ promotion of water infiltration along root channels and in soil, with better structural stability
○ control of runoff and sedimentation by stems and surface organic matter
○ increased organic matter and enhancement of soil productivity, encouraging greater vegetation growth and better protection of the soil
○ increased soil organic matter reducing soil erodibility
○ reduction in the need for costly conservation structures and other inputs
○ better production and financial returns, enabling further capital investment in land improvement.

Examples of how agrodiversity can improve resilience in the face of land degradation can be found within each of the principal elements of agrodiversity. The discussion in Chapter 3 is also relevant here, and in the case studies valuable material is presented in Chapters 14, 15, 22, 23 and 24, in particular.

Biophysical diversity includes the natural resilience of the soil, which is buffered by increases in organic matter. For example, Elwell (1986) has shown conclusively on high-clay Zimbabwe soils that the stability of water-stable aggregates is crucially dependent on maintaining an organic carbon level of 2 per cent or more. Below this threshold, aggregates break down quickly, surface sealing occurs, runoff accelerates and soil erosion results. Above this level, surface aggregates keep a good surface-water storage capacity, allow adequate infiltration and ensure optimal plant growth. Farmers identify the biophysical diversity in which they operate, and manage it in different ways. Resilience relies on production of sufficient biomass, having a range of different types and qualities of organic materials, and its reincorporation into the soil.

Management diversity is often founded on local knowledge. Indigenous and adapted techniques of soil and water conservation are the most immediate way in which this form of agrodiversity may control land degradation. For example, Bunch and López (1999) have described how local innovation based on the use of mainly biological techniques has transformed many Central American rural areas. In one small community in Honduras, 16 innovations have been added to the already biodiverse farming system: four new crops; two green manures; two new species of grass for use as contour barriers in vegetable fields; new chicken pens made of king-grass; marigolds to control nematodes; the beans *Dolichos lablab* and *Mucuna pruriens* to feed cattle and chickens; Napier grass to control landslips; and so on. An increasing number of studies now attempt to assess the knowledge that farmers hold and the results of their experimentation. An example is soil fertility improvement in Ethiopia by fallowing, manuring and use of crop residues (Corbeels et al., 2000). Management diversity in an agrodiversity framework expresses this wide variety of practices and innovations that use the attributes of many plant species. Such practices add resilience to the whole farming system, and they also increase production through conserving soil and water, improving soil fertility and controlling land degradation.

Agrobiodiversity in the form of intercropping is the single most direct way in which biological diversity can assist the control of land degradation. Having more plants in the field over longer time periods substantially increases the vegetation cover. Having a greater variety of plants enables species to exploit the many micro-ecological niches within a farmer's field, again maximizing vegetation cover. This cover has been demonstrated to be the single most influential factor in reducing erosion rates (Stocking, 1994). A monocrop of maize may typically have a mean annual vegetation cover of 30 per cent. Under the same soil conditions, an intercrop of maize–velvetbean (discussed as an example below) would have about 60 per cent. This doubling of cover reduces erosion hazard to one-sixth of the maize monocrop situation. Agrobiodiversity includes the many complex sequential and spatial intercropping practices employed by farmers, and the many systems of agroforestry that are so well exemplified in smallholder farms. It embraces multistorey cropping in some of the world's most complex and productive agricultural systems, such as the Chagga home gardens in northern Tanzania and the Kandy multistorey gardens in the hill lands of Sri Lanka. A good review of the many beneficial processes in agroforestry systems and how they directly affect the soil and control land degradation is that by Huxley (1999).

Finally, organizational diversity expresses the variety between farms and plots of land, and of the use of different resource endowments, income and labour sources. Tengberg et al. (1998) show how the characterization of farmers into different resource levels largely explains broad differences in practice and land management. The variables found most important in explaining farmers' engagement in local soil and water conservation practices were access to cash income, farm size, livestock ownership, labour availability, input use, cultivation methods and gender of head of household. Use of techniques of conservation such as trashlines, *Fanya-juu* terraces, log-lines and stone bunds, was found to be related to these variables in Mbeere District of Kenya. The farmer's actual choice of conservation technique is also related to type of soil and quality of land (biophysical diversity).

How agrodiversity promotes food security

A central aspect of livelihoods is food security. This is primarily a function of maintaining adequate production of food, and providing the variety of foods needed for a balanced diet. Implicit in this link between agrodiversity and land degradation is increased production. Unfortunately, supporting data for these benefits of land degradation control are elusive. Most commonly, the reverse relationship is reported, showing that land degradation *decreases* crop yields. Nevertheless, it is reasonable to assume that, if land degradation had *not* occurred, and if conservation were practised, then yields would be saved. It is also reasonable to assume that effective techniques of land degradation control enacted through agrodiverse strategies would also promote greater production and better food security.

Results from a large number of experiments sponsored by the UN Food and Agriculture Organization (FAO) in Africa and South America show the nature of the general relationship between erosion as a measure of land degradation and crop yield as a measure of soil productivity and food security (Tengberg and Stocking, 1997). The initial decline is steep but depends on the specific soil type, which is an aspect of biophysical diversity. However, actual decline in yields in a farmer's field is far more sensitively related to specific erosion rates. In terms of aspects that are amenable to farmer's management, erosion rates are primarily a function of vegetation cover. Farmers who use diverse methods and plant inventories are more likely to be able to sustain yields at reasonably productive levels with low inputs.

An example: the maize–*Mucuna* system in Latin America

What sort of agrodiverse practices provide for better cover and for meeting real livelihood benefits in smallholder farming systems in the tropics? Only one, albeit quite remarkable, example will be presented here. The maize–*Mucuna* (velvetbean) system has been gaining enthusiastic acceptance in many South and Central American countries.

On the Ferralsols (Oxisols) in the western hilly parts of Santa Catarina State in southern Brazil, many smallholders used to gain only a precarious living through arable cultivation of maize on these very erodible soils. Conservation programmes had largely failed because the cost of structures was greater than any benefits to be gained in the short term. The maize–*Mucuna* system has transformed both the erosion situation and the maize-based livelihoods of local farmers. To account for this change, take the stages in a seasonal agricultural cycle. First, maize is directly drilled by hand (or with a horse-drawn drill) into the soil. Two weeks later, as the maize has started to germinate, *Mucuna pruriens*, a climbing legume, is also directly drilled. The maize grows rapidly and is harvested. The *Mucuna* is slower to grow, reaching maximum growth rates only after harvest of the maize. By the early dry season, the *Mucuna* completely covers all the maize stalks with a dense mat of creeping biomass. Over winter the *Mucuna* dies back, forming a thick black mulch over the soil surface. In the second season, the farmer simply directly drills into this mulch for the new maize crop. The benefits are manifold:

○ an increased yield of maize – the *Mucuna* is largely non-competitive with the maize, and farmers relate how the benefits of the whole system translate to approximately tripled yields compared with monocropped maize
○ weeds are out-competed or completely stifled by the *Mucuna* mulch; this saves considerable labour at peak times
○ tillage is unnecessary, again saving labour and cost
○ at favourable sites, the *Mucuna* even sets sufficient seed to make replanting in future years unnecessary.

The maize–*Mucuna* system is an excellent example of how an intercrop (agrobiodiversity), managed in a way that avoids mechanical tillage (management diversity), using the substantial labour benefits of suppressing weeds (organizational diversity), while improving soil structure and chemical fertility (biophysical diversity), can transform rural livelihoods. This is a special case, which was promoted by missionaries in many South American countries, and further spread by farmers rather than by the extension system.

Conclusion

'Biological diversity' is a term of our time. Its protection in areas of land use is now one of the highest priorities on the international agenda as expressed by the government signatories to the CBD. Yet, biological diversity remains an enigma. It is difficult for non-ecologists to link the term with items of practical or economic relevance. Modern agriculture is the antithesis of biodiversity; and rural development in the poorer parts of the tropics would seem to have far more pressing problems than protecting obscure species and habitats. However, for the CBD to make headway, it is vital that biodiversity be seen in its pragmatic context, not only in ecologist's jargon as a 'life-support system' but also in a real practical context in making the environment more secure and people's livelihoods better supported. PLEC and its collaborating farmers are forging the link between global demands and local necessities.

Dynamism is the defining characteristic of these links. As Scoones (2001) points out in the context of soil fertility, issues of spatial and temporal dynamics, of diversity and difference, of history and change, of socio-economic setting and relationships, of policy context and trends, are central to a balanced analysis of what is happening in rural society. Agrodiversity captures most of these components of change, and its analysis attempts to focus on multi-disciplinary solutions to complex problems of environmental degradation and human poverty.

This chapter has focused on global issues and their relationship to the very local issues of sustainable livelihoods. It has only very recently been recognized that opportunities for action at local, national and international levels, through targeting agricultural biodiversity, can reduce poverty, promote sustainable rural development and improve food security (see Cromwell et al., 2001). PLEC has been demonstrating this in the countries in which the project works for some years, and this book now draws the attention of a wider audience to the achievement of the farmers in themselves making this link.

5 Guidelines on agrodiversity assessment

HAROLD BROOKFIELD, MICHAEL STOCKING
AND MURIEL BROOKFIELD

Introduction

RECORDING OF AGRODIVERSITY and biodiversity are best thought of as two dimensions of the one job. There is one central task that is common to both: the determination by observation and field collaboration with farmers of the broader Land-use Stages or types at landscape level, and the finer detail of Field Types at site level. We begin the chapter by examining the common basic task, go on to discuss issues of data collection and analysis, and conclude with a fairly extensive data collection checklist.

Land-use Stages and Field Types

It is important first to define the two principal terms used as the elements of managed and unmanaged landscape that are basic to sampling and description. Field Types are discussed in greater detail in this chapter, and Land-use Stages in Chapter 6. Good examples of how field scientists have presented findings on Land-use Stages and Field Types can be found in Chapters 14 (Tanzania) and 21 (China).

Land-use Stages

Land-use Stages are areas of broadly common ecology, land use (or its absence), and especially recent land-use history. Without detailed inventory, they look like one class of land use, with one class of land cover. They may cover up to several square kilometres in extensive systems, or smaller areas in more intensive, high potential management systems. We use the term 'Land-use Stages', but they are broadly comparable with the land-utilization types discussed in the FAO land evaluation literature (FAO, 1976). FAO (1983: 26) sets out a useful table of headings for describing land-utilization types, plus some suggestions for descriptive and semi-analytical quantification. The FAO methodology does not have the same purpose as ours, but it is an attempt to grapple with the same order of complexity. Some readers may find it useful.

Even where a land-use map is available, or can be generated from remote sensing imagery or photographs, transects in the company of farmers are an essential early step in the identification of Land-use Stages. Whether large or small, they should be recognized at a landscape scale, broadly at a map scale of about 1:25000–50000. In Chapter 6 the examples given are fields under annual (or semi-annual or longer-than-annual) crops, agroforests, fallows, orchards (including fuelwood plots and cash crop plantations), native forests, house gardens, and the edges between different types.

Land-use Stages are the primary sampling units for inventory of plant species, and they are the basic landscape level units for the analysis of agrodiversity. Zarin et al. (Chapter 6) use the term 'stages' to emphasize how one land-use type can be converted into another, both by successional processes and by farmers' own action. Over the four years of the project, stage transitions have been observed in all study areas.

Field Types

The distinction between Land-use Stages and the usually smaller Field Types is that the latter are specifically defined by farmers' practices, and not just by observation. This is the level of detail that farmers themselves recognize. It is the level where farmers assign vernacular names to soil types and where microclimates are employed as specific production opportunities. Field Types are assemblages of individual fields, managed sections of fallow or forest, agroforests and orchards, in which a similar characteristic set of useful plants is encountered, and in which resource management methods have strong similarity.

Recording should follow the farmers' own categories for management of diversity. Although each individual field is different, there is often considerable similarity over quite a large area. Commonly, farmers develop specific sets of Field Types, in each of which they recognize similar biophysical attributes, use similar management methods, and grow similar sets or combinations of crops.[1] There may be only one, or a large number, of Field Types within each Land-use Stage.

In some systems, these Field Types shift across the landscape from year to year. At a village in Amazonian Peru, Christine Padoch and Wil de Jong (1992) identified what they described as 12 distinct 'farming systems' – management associations that we would call Field Types – in one small community in the Peruvian Amazon. In 1985, among the 46 households, 39 ways of combining the 12 types were found. Many changed these combinations in the following year. These farmers were using the dynamic environment of a shifting flood plain, as well as the dry land above it.

Where land rotation is practised, formerly cropped fields leave behind them successional (fallow) management types from which crop plants may continue to be taken, and in which the successional vegetation may be planted or managed. These constitute a further set of Field Types. The 1999 progress report from the Amazonia Cluster contained a classification of fallows at Amapá, where they are of major importance as production spaces. Five field types were identified:

○ fallows in which vegetation was dominated by bananas, planted during the field stage and then managed in an agroforestry pattern to control losses from disease

○ fallows dominated by planted, broadcast or naturally regenerated açaì palms, and managed by protection for commercially valuable fruits and palm hearts

○ fallows in which the plant community was dominated by timber species that were protected or planted during the field stage, and were harvested as timber

○ fallows in which the vegetation was dominated by fruit species that were planted or protected during the field stage

○ fallows in which vegetation was not managed or enriched during the field stage, dominated by vines, shrubs and trees, and destined to become new field sites.

In other systems, Field Types change in response to processes such as land degradation, or external pressures such as market forces. At the demonstration site in Bushwere, Uganda, bananas and coffee dominated the lower slopes of the landscape. As urban demand for bananas has increased and coffee has become increasingly susceptible to berry diseases, Field Types have shifted, but only after a time lag occasioned by farmers' reappraisal of what was in their own best interests with these perennial crops. Meanwhile, the upper slopes have been exploited for their soil fertility through crops such as sweet potatoes. As soil quality declines, land use changes to cassava cropping and increasingly lengthy fallows. So, Field Types change over time and space. However, any assessment at this level also needs to recognize that Field Types in different Land-use Stages are linked. In the Uganda case, eroded sediment from the upper slopes was providing the opportunity for increasing the area devoted to the banana-based Field Type. Such aspects of biophysical diversity are included in Part B of the checklist appended to this chapter.

In systems where stage transitions take place infrequently, the Field Types are more permanent, and are often grouped within areas of broadly similar ecology. An example is the intensively cultivated and manured infield, versus the more extensively used outfield, common in the savanna regions of Africa and most sharply represented within the Fouta Djallon of Guinée (see Chapter 23). Another example is the division of land between seasonally irrigated terraced or ponded fields, wet fields fed only by rainfall, dry fields which are alternately cropped and fallowed, planted and managed agroforests, and very mixed home gardens. This repeated pattern of just five main types is commonly found in Yunnan, and occurs widely across southeastern Asia.

Fields also have edges, whether separating fields of different types or of the same type. At the field level, the edges may have a specific management role and a distinctive plant ecology. An example is discussed in Chapter 22 (Thailand). Thus live hedgerows and the risers separating terraces are edges. They may have a role in soil and water management as well as being used to provide or grow distinctive useful plants; some of these have the additional functions of fertility management or soil stabilization. At the smallest level, edges also include trash lines, small stone walls or low wooden fences. While not all these smaller features are significant from the point of view of plant bio-diversity, they are significant from the point of view of resource management.

Field Types, bringing together crop selection and resource management in distinctive ways, often arise in response to specific ecological conditions. While specific ecological niches may be used in specialized ways, these ways tend to be repeated over a large area. Field Types are also the means by which farmers most effectively mobilize their labour and allocate their resources. In many areas of the world, the basic reason why repeated patterns of Field Types come into existence would seem to be that they simplify work routines, and the problems of daily decision making.

Notwithstanding the enormous internal diversity of cropping patterns, it is quite common to find the land used under only a small number of basic management systems, even across significantly different ecological zones. To recognize them necessitates not only repeated observation, but also the co-operation of the best and most alert farmers. It is easy for observers not trained to look for micro-features in the managed landscape to miss a great deal of relevant detail.

Recording diversity within the elements

Once the sampling or selection frame, in the form of identification of Land-use Stages and especially of Field Types, has been done sufficiently for detailed work to begin, observers need to seek a range of information. Discussion of what is needed can best be classified within the four elements of agrodiversity discussed in Chapter 2. This review is followed by a brief discussion of analysis and presentation, and then by a checklist to assist in recording data in selected or sampled fields (the site), and on the organizational diversity of the farming households that operate these fields. The most important elements in using whatever recording design is most appropriate to the area are to be logical and consistent, and to be able to relate detailed work in the fields to the characteristics of Field Types and Land-use Stages at the landscape level. Appropriate map scales would vary greatly from area to area, but for the landscape may be in the range of 1:25 000–50 000, and for the site 1:5000–7500. Particular areas within the site may need to be sketch-mapped at a larger scale.

Experience has shown that drawing a field sketch map with the farmer has many advantages that can be used later when detailed assessments are made.

Making a participatory sketch map is a good 'ice-breaker' – if the farmer is positively engaged with an output from the field visit which he or she can keep afterwards, then that alone brings the farmer on to the professional's side in a shared experience. It immediately identifies the Field Types recognized by the farmer as being distinctly different by virtue of their biophysical condition (slope, soil) or their actual or planned land use (home garden, perennial crop plot, grazing paddock, etc.). It provides a baseline from which to organize the detailed assessments for each Field Type.

Ultimately, sketch maps for use by observers in the field, at different scales according to the amount of detail investigated, might show:

○ areas having similar vegetation
○ observations on key biophysical aspects, such as slope, soil, rocks, springs, and drainage
○ all farmed land and, where feasible, Land-use Stages and Field Types characterized by similar forms of management and cropping patterns
○ settlements and roads, and other basic features
○ the outline of any area (or areas) studied and mapped in greater detail
○ the position of biodiversity-inventory sample quadrats, and soil-sample sites.

This needs to be complemented by description, and by background material for the population of the whole landscape area. A suggested list of items of importance are indicated in the Agrodiversity checklist towards the end of this chapter (Part C, Organizational diversity). Site maps, covering much smaller areas, need to carry greater detail, but for reasons of protecting farmers' intellectual property rights we advise that for publication purposes, the actual location of the sites be shown only in an imprecise manner.

Selection of sample fields: looking for the unusual

Random sampling, insisted on only for native forest, is distinguished from biased sampling in Chapter 6. Biased sampling is appropriate in all other Land-use Stages if observers seek data representative of the most productive or species-rich examples within a Field Type. With management diversity as well as agrobiodiversity in mind, we introduce a further reason for biased sampling. As discussed at several points in this book, especially in Chapter 9, PLEC has sought expert farmers with whom to work in demonstration sites. These are 'farmers who put their expertise into patterns that combine superior production with preservation or even enhancement of biological diversity in their fields' (Padoch and Pinedo-Vasquez, 1998: 8). It is not easy to identify such farmers, but one good place to do so is in the fields themselves. We are certainly concerned with what the generality of farmers do, but also look for the unusual and innovative in resource management techniques, to sample such fields where they are found, and get to know the expert farmers.

One way of selecting sample fields within a Land-use Stage or Field Type would be walking and briefly recording diversity along intersecting short transects designed to take different Field Types into account. Field workers need to stop frequently if important detail is not to be missed. A lot of information arises from careful observation, and from in-field discussion with farmers. It is important also to ensure that fields are chosen to be representative of each Field Type. It is in this process that Field Types are likely to be subdivided, as described in the next chapter.

Recording of management diversity in sample fields can accompany the recording of biodiversity within the cultivated and fallow areas. Especially if two or more field scientists are present together, work on management diversity of selected fields can be combined with biodiversity inventory in the quadrats. The whole field and its edges, not the quadrat, is the appropriate unit for recording management methods, but the tasks can nevertheless be conducted at the same time, thus minimizing interference with the normal activities of cooperating farmers. The farm operator, whether owner or tenant of the field, should always be present, and his/her name recorded in order to cross-check with the data obtained on organizational diversity.

Recording biophysical diversity

The recording of biophysical diversity at site level should also involve the farmer. The aim is not to produce just a technical land resource survey, although there may be a need to collect samples of soil, vegetation or water for laboratory analysis in order to provide scientific explanation for farmers' differentiation of aspects of Field Types. Recording should also focus on change, such as how soils have been 'manufactured', or how supposedly natural forests have been managed and plants domesticated. Land degradation is one particularly important aspect of biophysical diversity. In one part of the landscape, perceived land degradation may give severe restrictions to the productivity of Field Types, whereas in another part of the landscape the products of up-slope degradation may provide excellent inputs of additional water and nutrients.

The soil is usually the best integrator of the wide array of possible biophysical measurements, and it is also an aspect with which land users have greatest familiarity. Our Mexican colleagues, for example, at first used Spanish language names for soils when talking to the farmers, and the farmers were happy to go along with these as the name was usually the colour of the soil. However, it was only after starting to use local-language (Massawa) names that a whole new area of management diversity opened up for inspection by the scientists. Immediately, farmers could relate how they used particular types of rock deposits to improve fertility, with explanations of how this soil biophysical diversity largely controlled where they would plant local maize varieties.

Often farmers also use vegetation indicators. The occurrence of bracken ferns in fields in the mid-hills of Nepal is an indicator used by farmers to show the history of land use and current poor soil quality. Similarly, the indigenous fruit tree of Zambia and Zimbabwe, *Parinari curatelifolia*, indicates the occurrence of sodic soils and associated problems of dispersion of clays and accelerated soil erosion. Farmers know this well, and the field assessor will thus gain considerable understanding of the important dynamics of biophysical diversity.

Recording organizational diversity

Organizational diversity differs from the other elements in that it cannot be recorded except at the level of whole farms compared with one another. Farm layout is an element potentially capable of being mapped, but recording of other aspects calls for repeated discussion with farmers. So far as is possible we advise against formal interviews with farmers whom researchers grow to know well. Investigations of the variable resource endowment among farm households need to begin at community level. If the PLEC model of selecting expert farmers as primary partners in demonstration site work is not followed, sample (or contact) farmers should be selected only after classes of farmers having different resource endowments have been determined.[2]

As discussed in Chapter 3, land tenure can be a particularly important variable, as it can have important consequences for land management and agrobiodiversity. The conditions of land tenure should be carefully recorded. These data should be complemented by information on the population of the landscape area, including its demography, migration history, form of social organization, and arrangements for marketing of produce. In turn, this nests into wider-area information on the regional and national economies, policies and political forces.

Recording findings: different ways

One basic way of recording diversity would be to prepare a matrix in which different sampled fields within Field Types are treated as units within each of which particular methods, and crops, can be simply recorded by their presence or absence. Such a matrix can be prepared in note form in the field then transferred to a data-organizing system, among which the simplest to use is probably an Excel workbook. Another way is to use a roughly surveyed map, where areas can be measured and data on one aspect can be compared with another using GIS. Both methods have problems of which the field workers need to be aware.

Tabulated data can readily over-simplify complexity. Data tabulated or presented on maps have to be divided into classes, creating a false impression of uniformity over tracts of land sharply distinguished from other tracts of land. There needs also to be description, as in the examples presented in Chapters 14 and 21. The sample quadrats used for biodiversity inventory are free from these problems, but they are not appropriate for analysis of management diversity, where the unit should be the field. Data obtained only from within or close to quadrats may omit important features that lie outside their limits, but have a role in relation to what is observed within them.

Data on organizational diversity need to be related to observation and recording in the fields. Fields therefore need to be given numbers, related to the numbers given to the farms. We do not suggest that all the fields of any farm need to be sampled, but it may be important to record the number of Field Types that are represented within a selection of farms. Such a selection should include farms of the expert farmers who are identified, but it should also include farms of both well-to-do and poorer farmers. One constraint to diffusion of good practices would thus be identified. Any wealth ranking of farmers is therefore a job that needs to be done at an early stage in the work.

A database can be only a partial product of the whole work, and a range of ways of presenting agrodiversity is necessary. It is probable that different combinations of methods will best suit different areas and their demonstration sites. What is presented here is only one way of going forward. The information discussed below is needed for work at the site level of particular fields and farms. It is simplest to present it in checklist form.

Analysis and discussion of findings

The different elements of agrodiversity (agrobiodiversity included) need to be explained in relation to one another, and used in relation to one another in further work with the farmers. Statistical measures of diversity are an end-product of biodiversity analysis, but have few parallels in dealing with agrodiversity. It will certainly be valuable to obtain measures of crop plant diversity, and it will be of particular significance to obtain these at the level of local (or landrace) varieties. Elements of the biophysical environment, such as soil fertility, and of the comparative status of farming households, can be reduced to simple statistics.[3] However, the core of agrodiversity lies in management, and no statistical indices yet exist for the analysis of resource management practices. Here we are dealing with what is technically termed non-numerical unstructured data. There are various computer routines for dealing with data of this nature, indexing them and finding structure within them, but some are very consuming of computer space.

For reporting, the results of agrodiversity analysis need to be presented in such a way as to exhibit the depth of variation present, but without overwhelming the reader with detail. It will usually be best to determine, before writing, what are the main organizing principles in the situation described. This usually means placing the results in a regional or national context, and in the context of the driving forces of change in the recent history of the area.

For example, an area might have been settled by its present people, after moves from elsewhere, only within the past 50 years. There may have been major changes in economy and politics during the lifetime of people still active, impacting the nature of decision making and, very importantly, the conditions of land tenure. Whether by natural growth or immigration, or both, there may have been a major increase in population. Cash production, and other forms of commerce, may have become dominant only in recent years. There may have been important environmental changes due to shifting rivers, soil degradation or recent deforestation. Whatever is most relevant should be the peg on which discussion hangs, and around which data analysis can be focused.

The same considerations that affect land use also affect agrobiodiversity. The explanatory elements of agrodiversity discussed in this paper are also explanatory elements of agrobiodiversity. There is no one single formula for all areas. The job of reporting is an iterative one. It begins with the presentation of data. As work goes forward, and as the researchers and the farmers gain increased familiarity with the aims and interests of one another, the data will improve. It will then become easier to interpret the data, both for presentation and most importantly in order to design further work with farmers.

An agrodiversity checklist

What follows is a suggested checklist only, for use in sampled or studied fields within Field Types, and for recording the organizational diversity of the farms operating these fields. Its purpose is to help ensure that comparable data are collected. Not all of it will be applicable in all areas, and other items will need to be recorded in some areas. The main common requirement is that information should be collected and recorded under each heading (shown in bold type) within each category of agrodiversity, to enable early identification of what is important and identification of aspects that are worth investigating further. We suggest that if you are creating a presence/absence matrix you use these categories. The indicative nature of the list outlined below is stressed.

Part A: Management diversity

The unit is the whole field and its edges, not a quadrat within it. As chosen fields are representative of Field Types, it is important to look also at the

environs of the selected fields in order to ensure that nothing of importance has been missed. Similar, but not identical, lists would apply to information needed for sampled agroforests and home gardens, and their edges.

Site preparation

Includes: tree-felling, slashing, burning (whole field, patches, debris only), clearance of preceding crop, ploughing of whole site ahead of any planting.

Methods of field surface preparation

Includes: holing, tilling, mounding, ridging, and all other ways of preparing the ground for planting. Tools used. Measurement of mound height, ditch depth, and cover depth if mounding, ditching, burying of compost or of cut or uncut grass are practised. Observations need to be made at the time land is being prepared for the crop.

Major land management

Includes: construction of terraces, walls, with details of height, construction material and method, slope of terraced land. This also includes drainage and irrigation extending over substantial areas. It overlaps with biophysical diversity in that the manufactured soils that are produced are products of major land management. Diagrams can be an important method of additional record.

Minor land management and soil or water retention

Small slope retention devices, of wood, stone or earth, within the field or on edges. The planting of soil-retention crops on steep patches. Soil drains are a feature. Barriers such as grass strips, soil or trash lines, log lines, stone lines, contour ridging, live barriers along the contour, and soil-retention fences are included. The dimensions, spacing, length of life, and replacement are important. Note whether they are traditional (meaning not recent introductions) or modern.

Planting material

For seed: source (own farm, other farmers, bought or supplied). For non-seed plants: nature of planting material and source as above. Planting methods: dibbled, sown in rows, inserted in holes, broadcast, transplanted. Note any volunteer crops that arise spontaneously without planting. Crop varieties where relevant: note local names for varieties, and list distinguishing characteristics.

Cropping patterns and rotations

To supplement information obtained in agrobiodiversity quadrats. Monocrop or intercrop: note the main crops and their planting schedules – all together, early or late. In order to supply an indication of crop rotations or sequences, record the previous season's crops and intended next season's crops. Crop-plant spacing (along and between rows) and densities. Distinguish planting around or on field edges, and plot edges.

Weeds and weeding

Weeds: severity of infestation, effect on yield, recent or invasive. Most serious weeds. Frequency of weeding, method (manual, hoes, chemical, other), time taken to weed one field. Proportion of weeds not removed (if any). Note separately weeding methods in agroforest plots and in home gardens.

Pests and diseases, predators

Principal pests, within soil, on surface, birds. Severity of pest infestation. Farmer's methods of control. Disease problems: nature, recency, severity, methods of control, if any. Predators: if a problem, methods of protection.

Harvest, processing and storage

Time of harvesting of different crops, both those harvested at one time, and those taken as needed. Methods of harvest. Labour requirement. For grains, note if threshed in the field or at home. For any crop, post-harvest processing, place and method of storage.

Livestock (see also under 'Organizational diversity')

Note if livestock are tethered in the field or on edges, or if allowed freely into the field at any stage.

Soil fertility maintenance

It is important to note farmer's assessment of fertility of local soils, and effect of different crops in depleting soils. Fertilizers: type, quantity, method, timing. Compost: content, how made, quantity, timing. Mulch: sources, quantity, timing (note effect on pests). Use of nitrogen-fixing plants in intercrops, rotations, quasi-rotations, or agroforestry patterns (overlaps with recording of agrobiodiversity, but needs a separate note where it is a conscious management practice).

Woodlot management

Includes: species grown; management (weeding, coppicing, felling), frequency of harvest. Presence or absence of any intercropped useful plants (see also under 'Organizational diversity').

Fallow management (agronomic aspects)

Farmer's management: enrichment planting, weeding, slashing, elimination of unwanted species, harvesting of useful products, trapping and hunting wildlife. This aspect of management is very important in some areas.

Management aspects of otherwise unused land (native forest)

Includes: reservation of forest for watershed protection, sacred groves, burial places, use for trapping and hunting wildlife and harvesting of useful products, including wild vegetables and wood.

Part B: Biophysical diversity

The unit where measurement is required is the whole sampled or selected field and its edges, but it is important also to obtain observational data (only) from the whole landscape area within which Field Types are identified.

Physical features (whole landscape-level area)

Includes: the whole area of the studied tract of land, altitudinal range, types of landforms (e.g. plateau, scarp, valley floor, rock outcrops). Slope: per cent steep, medium, low, flat. Note any flood-prone areas (normal or exceptional), swamp.

Physical features (sampled field)

Note area, dimensions, altitude, slope, degree of modification by farmers (from 'Management diversity').

Soils

Soil samples should be taken in sampled or selected fields within each Field Type, and preferably at points used for the biodiversity inventory. Sampling should extend down to at least the base of the topsoil.

For each sample site, the following observations need to be made and recorded.

Assessment in the field

Soil type (local name; correlate with FAO/ UNESCO, Soil Taxonomy and/ or national survey classification later); soil texture (local descriptors); soil depth (both of topsoil and subsoil); physical properties (e.g. hard when dry); stony; drainage and infiltration (poor, imperfect, good; and note seasonal differences); workability (good, poor; and note seasonal differences); colour (description, correlate with Munsell Colour chart); local assessment of relative fertility and associated indicators; other definable field characteristics (e.g. micro- and macro-fauna). Test the infiltration rate.

For later laboratory assessment

A minimum list: soil texture; soil chemical fertility status (cations Ca, K, Mg, Na, and total N and available P); pH; base saturation; cation exchange capacity; organic matter status (per cent); C/N ratio; soil physical status (bulk density; available water storage capacity).

Erosion, degradation

Note: evidence of sheet erosion (micro-pedestals); tree mounds and plant root exposure; in-field bare patches; poor crop growth; soil accumulations against trees, fences, other barriers; armour layer (small surface stones); rills and gullies (dimensions, spacing; evidence of recent soil losses); other *in situ* soil degradation (surface crusting, waterlogging, salinity, sodicity, etc). Landslips, landslides and major earth movement.

Manufactured soils

Within the above, some soils are greatly modified by human use. They include soils that have been heavily manured or composted over a long time, with deeper surface horizons, usually higher pH and more available P, and soils created behind terrace walls. The fact of human modification emerges from information and observation. The characteristics to be examined for each sample taken are the same as above.

Microclimates

Check whether farmers are aware of specific contrasts in microclimate between different parts of their land, or of the whole landscape-level area.

Indicators

Note the indicators farmers themselves use: for soil quality, soil depth, planting position, yield levels, crop mixtures. The use of the same indicators and local-language names by the assessors gives a common language for use in interacting with land users.

Part C: Organizational diversity

The unit is each whole farm within which work is undertaken. Organizational aspects at community level are also important, for example in arrangements for resolving land tenure disputes, or managing ingressions into the land of others, for fuelwood or useful plants. The nature of these arrangements differs greatly between areas, and no checklist of these is offered.

The list is to help ensure that information obtained on farms is comparable and complete. This is not a design for any sort of formal questionnaire.

A few **census-type questions** should be asked. For the farmer, male or female, and for other adults present they include: name; age; sex, born here or elsewhere. For the farmer; are they head of household or not. If a migrant; from where and when. Years of education. Number of children, age and sex, present/absent.

(This basic information can be enlarged. Those with appropriate social science training may find it valuable to set out the relationships of household members in diagrammatic form.)

Other topics for record have more to do with the farm

Land tenure
Farmer gender. Ownership history of the farm. Whether the land has been subdivided, when, and approximate area of subdivisions.

Tenancy, formal and informal
If the farmer is a tenant, note if a share of crop is used for rent. Note use by any other household of part of this farm, with details of any crop share received. Check whether farmer or his/her household uses any part of another household's land, with details.

Crop and tree ownership
Any allocations to particular household members for their use (note particularly gender allocations). Note if any trees belong to someone who is not a member of the household, and if this farm household, or its members, own trees that are not on their land, with details.

Land-use history
Note that information on this topic is unlikely to emerge in a single conversation, and is likely to come forth only over long period of time. Topics for investigation: Length of time the farm has been in existence; the earlier history of the land; the main changes in land use/crops during the farmer's lifetime; land degradation, if any, during this time.

Land-use intentions
As with the previous question, this is not a topic on which information is likely to emerge in a single conversation. The farmer's intentions for the

future use of their land, both in the short term of the coming two years, and over a longer period.

Livestock
Details of livestock, large and small, that the household owns. How the livestock are fed. The products that the livestock contribute, and where they are housed at night (this question has cross-cutting relevance also to 'Management diversity').

Off-farm employment
(This is important in relation to farmers' commitment.) Check which members of the houshold have worked elsewhere during the past year and how much time they have spent away. From this, follow up with questions concerning how much the farm household depends on off-farm income. (Specific topics are not listed, as this is a very open-ended area.)

Food security
Ascertain whether the farm provides most (say, two-thirds or more) of the household food supply during the year, and if not, how much. Check whether there is a part of the year during which the farm does not provide sufficient food. If there is a bad season, and a crop fails, what does the household do?

Water
Note whether there is an adequate supply of water within easy reach and, if not, how far away is the collection point. The source of water: piped supply, wells, streams, tanks, other. Reliability of the water supply, and if it ever gets polluted or contaminated.

Fuelwood
(see also 'Management diversity'). Ascertain if the household has access to adequate supplies of fuelwood from its own land or community land. Ascertain if a part of fuelwood needs has to be met from native forest, including reserved forest areas.

Labour supply
(This important aspect needs to be explored over time.) Contributions of the family to labour on the farm; age group and gender of the contributors. Tasks done only by adult men; done only by women; also done by children. Details of any hired labour, and the season and work entailed. Check whether there is cooperation between farms for specific tasks.

Transport
Note any transport used to get to and from the fields, and of what type (include animals). Time taken to get to the furthest fields; distance from the farm to the nearest road. Record if the household owns its own pack animals, any bicycles, carts, motor cycles, or wheeled vehicles.

Marketing

The distance to the nearest (or most used) market; availability of transport to get produce to the market. Details of what is sold in the market, who does the marketing, and how often. Note if anything is sold directly from the farm.

Other aspects

To assist in interpretation, it may be useful to have a final checklist of farmers' views on constraints to their farm productivity which are outside their control. Many will have views on these topics. Perceived problems may include: lack of capital; need to borrow money at high cost; lack of security of land tenure; too little land; not enough good land; declining quality of land; lack of sufficient water; lack of sufficient livestock; inadequate supply of labour; poor marketing facilities; too many people (overpopulation); lack of community support, poor community spirit; too much intervention from the authorities; insufficient support from the authorities; pricing policies to the disadvantage of farmers.

Final note

The above is not an exhaustive list and much else could be added. These are aspects on which researchers need to be informed in order to work with the demonstration site farmers, both the identified experts and other farmers who cooperate in the work.

6 Guidelines on the assessment of plant species diversity in agricultural landscapes

Introduction

THIS CHAPTER PROVIDES definitions of essential agrobiodiversity terminology, a set of fundamental principles and practical guidelines for the collection of core plant species diversity data, and instructions for the reporting of those data. Analysis is discussed in a preliminary manner, and reference should be made to Chapter 8, where analysis is reviewed in greater detail. The information in tables and figures may be used in the field as a 'recipe' for collection of data. The text provides supplementary information and explanations. Our guidelines draw heavily on the varied experiences of the Amazonian, Chinese and West African areas represented by the authors.

Biodiversity surveyors must acquire significant familiarity with the landscape area before they begin to collect any core data, which require stratification of sampling based on Field Types. Methods for acquiring that familiarity are discussed in Chapter 5. In addition to stratification, the fundamental principles emphasize prioritizing sampling toward Field Types with high species richness, replication of sample areas, and collection of data within fixed sample plots which are remeasured at appropriate intervals in order to capture the critical temporal component of agrobiodiversity. The practical guidelines emphasize criteria for sample area selection, numbers of replicates, plot sizes, kinds of data to record, and sampling frequency. The suggestions for preliminary analysis include simple metrics for calculating the similarity in species composition among sample areas, and the development and uses of species–area and abundance–diversity curves.

The authors were brought together as the PLEC Biodiversity Advisory Group (PLEC-BAG) in July 1998. The primary purpose of PLEC-BAG was to ensure that the core PLEC agrobiodiversity data collection, analysis and organization were replicable, comparable between different areas, and of sufficient quality to meet international scientific standards. Meeting such standards was essential if the results of our work were to be accepted as valid and considered to have wider applicability.

Our specific tasks included recommendation of common guidelines for the collection and analysis of core agrobiodiversity data across different Clusters, recommendation and development of a database system for the organization of core data, and advising individual Clusters on agrobiodiversity issues as required. The core data refer to data that we anticipated including in the biodiversity database. This chapter is concerned with the first task, and is an outcome of the first full meeting of PLEC-BAG, which was hosted by the China Cluster in January 1999. A number of individual Clusters had already conducted a significant amount of agrobiodiversity data collection and analysis before these guidelines were written. While these data had considerable value, they could not be entered into a database.

Definitions of agrobiodiversity terminology proposed by PLEC-BAG and used throughout this chapter are presented in Table 6.1. Those terms most directly relevant to the collection of core species diversity data include Land-use Stage, Field Type, sample area, sample plot, nested plot/subplot, species abundance, species richness and sampling frequency. Additional terms most directly relevant to analysis of those data include evenness, species diversity (*sensu stricto*), and Sorenson's Similarity Index.

Fundamental principles

One of our first steps was to develop a set of fundamental principles for the collection of core plant species diversity data. A significant degree of familiarity with the landscape area around a demonstration site is required prior to the classification of Field Types and the collection of the core data. Methods for acquiring that familiarity, including farmer interviews, meetings and farmer-assisted transect surveys, are discussed in Chapter 5. Plant species diversity data collected during that familiarization process are extremely valuable and have a variety of uses; however, those data do not constitute core data as defined here. The fundamental principles for the collection of the core data emphasize the following points:

○ At each demonstration site, selection of sample areas for data collection must be stratified, based on Field Type. Table 6.2 gives examples of Field Types we have encountered within seven reasonably distinct Land-use Stages.
○ Selection of Field Types for sampling should prioritize those that appear to contain the greatest variety of species (high species richness). Researchers should pay particular attention to edges and other transitional areas that often contain high species richness but are generally ignored in sampling schemes because they do not fit neatly into any predetermined category.
○ For each selected Field Type, sample plots must be surveyed in multiple sample areas. Table 6.3 includes guidelines for the minimum number of replicate sample areas needed for Field Types within each of our seven Land-use Stages.
○ Data collection must occur within sample plots of fixed or measured dimensions. Guidelines for appropriate plot sizes and for the use of nested subplots to facilitate the sampling of understorey vegetation are included in Table 6.3.

Table 6.1 Definitions of agrobiodiversity terminology proposed by the PLEC Biodiversity Advisory Group

Term	Definition	Example	Source
Land-use Stage	A general land-use category based on vegetation structure and requiring a plant species-diversity sampling strategy distinct from that of other such categories	House gardens	This chapter
Field Type	A specific land-use category which corresponds to the finest-scale land-use division made by farmers and researchers	Monocultural cultivation of *Cassia siamea* in coppiced fuelwood groves in Yunnan, China	Brookfield, et al. (1999)
Sample area	A contiguous parcel occupied by one Field Type and selected for data collection	One selected *Cassia siamea* fuelwood grove	Avery and Burkhart (1983)
Sample plot	The portion of a sample area from which data are collected	A 20 × 20 m section of the sample area described above within which tree species abundance data (see below) are collected	Avery and Burkhart (1983)
Nested plot/ subplot	A smaller sample plot located within a larger sample plot	A 1 × 1 m section of the sample plot described above, within which herbaceous species abundance data (see below) are collected	Avery and Burkhart (1983)
Sampling frequency	The number of times a sample plot is measured	Annual sampling of citrus plantations in Ghana	Avery and Burkhart (1983)
Species abundance	The number of individuals of a species present within a sample plot	112 *Euterpe oleracea* stems in a house garden plot in a site in Brazil	Gove et al. (1996)
Species richness	The number of species present within a sample plot or a larger unit of analysis	90 species present within all of the sampled house gardens in a site in Brazil	Gove et al. (1996)
Evenness	The equitability of species abundances within a sample plot or larger unit of analysis	High evenness is where numbers of individuals are equitably distributed among the species present	Gove et al. (1996)
Species diversity	Any of a number of statistical properties which describe the relationship between species richness and evenness within a sample plot or a larger unit of analysis	The Shannon (or Shannon-Wiener) Index	Gove et al. (1996)
Sorenson's Similarity Index (S_s)	$S_s = 2Tc_{i,j}/(T_i + T_j)$ where: T_i and T_j = the number of species in sample units i and j, respectively and $Tc_{i,j}$ = the number of species common to sample units i and j	41 per cent similarity between two house gardens from a site in Amapá, Brazil	Jongman et al. (1995)

Table 6.2 Land-use Stages and Field Types initially identified by the PLEC Biodiversity Advisory Group

Land-use Stage	Examples of Field Types		
	Amazonia	China	West Africa
Annual cropping	Sugarcane	Wet rice field	Maize, millet and cassava monocrops
Agroforests	Banana and maize intercrop	Rubber, passionfruit, upland rice intercrop	Mixture of annual crop species with trees
Fallows			
Grass-dominated	Recent abandoned pasture	*Chromolaena* spp.	Along the Sekesua transect, Ghana
Shrub-dominated	Early regrowth following agricultural abandonment	Eight years after sugar cane abandonment at Baihualing	On the Accra plains
Tree-dominated	*Calycophyllum spruceanum* stands	Rare due to fuelwood harvest	Abandoned agroforest
Orchards	Banana plantations	*Quercus acutissima* coppice fuelwood stands	Oil palm and coconut plantations
Native forests	*Várzea* forest	Nature reserve forest	Sacred grove forest
House gardens	*Euterpe oleracea*-dominated garden	Extremely varied, often high in endemic species	Common around homes
Edges	Banana or annual crop boundary	Community forest–sugarcane boundary	Not yet surveyed

○ To capture the temporal component of agrobiodiversity, sample plots must be remeasured at appropriate intervals and for this purpose they must be physically marked. Table 6.3 also includes guidelines for sampling frequency.

Land-use Stages and Field Types

On the basis of vegetation structure, Land-use Stages are defined as distinct categories requiring different sampling strategies. We initially identified seven Land-use Stages: annual cropping, agroforests, fallows, orchards, native forests, house gardens and edges. We further divided fallows into grass-dominated, shrub-dominated and tree-dominated subgroups. Table 6.2 provides examples of these Land-use Stages from Amazonia, China and West Africa. We recognize that there may be additional Land-use Stages not included in Table 6.2 in areas with which we are unfamiliar, and that the Land-use Stages are not as discrete as their separate listing suggests. There may be overlap among several stages, particularly between fallows, agroforests, orchards and native forests. We do not believe that overlap will have a significant influence

Table 6.3 Data collection guidelines recommended by PLEC-BAG for core species diversity data for different Land-use Stages

| | Annual cropping | Agroforest | Fallows | | | Orchards | Native forest | House gardens | Edges |
			Grass-dominated	Shrub-dominated	Tree-dominated				
Sample area selection	Random or biased	Biased	Random or biased	Random or biased	Random or biased	Random or biased	Random	Biased	Biased
Minimum number of sample areas	3 within each Field Type	3 within each Field Type	3 within each Field Type	3 within each Field Type	3 within each Field Type	3 within each Field Type	5 within each Field Type	10	10
Plot size	1 × 1 m or 5 × 5 m	5 × 5 m or 20 × 20 m with nested 1 × 1 m subplots	20 × 20 m with nested subplots	20 × 20 m with nested subplots	20 × 20 m with nested subplots	20 × 20 m with nested subplots	20 × 20 m with nested subplots	Entire sample area with nested subplots	Sample in 1 m² increments
Minimum data to record	Presence, abundance & utility	Presence, abundance & utility	Presence, abundance & utility	Presence, abundance & utility	Presence, abundance & utility	Presence, abundance & utility	Presence, abundance & utility	Presence, abundance & utility	Presence, abundance & utility
Sampling frequency	Seasonal in year 1; once in each of the following 2 years	Seasonal in year 1; once in each of the following 2 years	Seasonal in year 1; once in year 3 only	Seasonal in year 1; once in year 3 only	Seasonal in year 1; once in year 3 only	Seasonal in year 1; once in year 3 only	Seasonal in year 1; once in year 3 only	Seasonal in year 1; once in each of the following 2 years	Seasonal in year 1; once in each of the following 2 years

on the practical utility of the Land-use Stages for distinguishing species–diversity data collection strategies. The inclusion of edges as a separate category is intended to encourage their sampling as discrete units.

In ecological terms, a Land-use Stage is analogous to a successional stage. We chose the term to reflect the dynamism we have seen, where some stages can be rapidly converted into others through very active or sometimes relatively passive management. In Yunnan, China, we have seen the conversion of centuries-old wet-rice terrace land to house gardens as a response to changes in markets, tenure and governmental policies. In Ghana, the conversion and abandonment of a number of traditional and industrial planting systems have led to frequent Land-use Stage alterations. In Amapá, Brazil, we have documented very rapid transitions among virtually all Land-use Stages.

The definition of Field Types is necessarily an iterative process. Researchers should expect that the number of Field Types will grow as familiarity with farming systems and with the landscape increases, and as new Field Types are added and others divided. In a few cases, a particular Land-use Stage may contain only one Field Type; house gardens are sometimes an example of this, particularly where house garden production is focused on one major cash crop (e.g. açaí fruit at the Amapá demonstration sites in Amazonia).

We view the relationship of Land-use Stages to Field Types as analogous to that between an ecological community and the niches contained within it. Modern ecological theory conceives of the niche as an *n*-dimensional space (Whittaker, 1975), defined by a very large number of biotic and abiotic variables and their interactions; given the multivariate determinants of Field Type discussed in Chapter 5 it seems appropriate to view it similarly.

Data collection

Table 6.3 presents our recommendations for sample area selection, the minimum number of sample areas required, appropriate plot sizes, recording of species richness, abundance and utility data, and sampling frequency. Each of those five issues is discussed below.

Selection of sample areas

Selection of sample areas within a Field Type may be either random or biased, depending upon the observers' goals, and it is important to realize that opting for random or biased sample area selection has important consequences for the interpretation of results. Random selection of sample areas within a Field Type may be accomplished following the development of a detailed land-use map of the demonstration site, which would necessarily include Field Types as mapping units. Random sample area selection is appropriate if the Cluster wants to collect data representative of the Field Type within the area as a whole. Biased sample area selection is appropriate if the observer wants to collect data representative of the most productive or the most species-diverse examples within a Field Type. Biased sampling can often be accomplished through a combination of farmer interviews and selective visits to sample areas identified as unusual by the farmers themselves. Biased samples are not representative of the average for the area as a whole, but may be useful representations of the unusual or the exceptional.

Within any given area, it may be appropriate to select sample areas randomly for some Field Types and to bias sample area selection for others. We recommend random selection of sample areas for native forest Field Types, and biased selection of sample areas for Field Types within the agroforest, house garden and edge Land-use Stages. Where highly managed fallow Field Types are present, biased selection is recommended; otherwise, fallow Field Types may be sampled randomly.

Minimum number of sample areas

Replication of sample areas is distinct from replication of plots within a sample area; the latter, which has been characterized as 'pseudo-replication' by Hurlbert (1984), may not be used as a substitute for the former. We recommend a minimum of three replicate sample areas for Field Types within annual cropping, agroforest, fallow and tree-crop Land-use Stages. A minimum of five replicates is recommended for native forest Field Types, and at least ten replicates should be used for house garden and edge Field Types. These are minimum guidelines only, and were selected, based on the experience of BAG members, to reflect the amount of replication generally needed to adequately represent within-Field-Type variation in species diversity. Species–area curves, discussed below, are useful guides for estimating when sufficient replication has occurred.

Plot size

Three plot sizes were selected as standard frames for the collection of core species diversity data: 1×1 m, 5×5 m, and 20×20 m. The 1×1 m frame may be appropriate for sampling some Field Types within the annual cropping stage and as a nested subplot for sampling the herbaceous layer of Field Types within the agroforest, fallow, orchard, native forest and house garden stages. As appropriate, the 5×5 m frame may substitute for or be used in conjunction with the 1×1 m frame; the 5×5 m frame may also be sufficient as the basic unit for sampling some Field Types within the agroforest stage. The 20×20 m frame is appropriate for use in agroforest Field Types characterized by wider spacing, and as the basic frame for sampling Field Types within fallow, orchard and native forest stages. Within the grass and shrub-dominated fallow substages, we recommend establishment of the 20 × 20 m frame even if only nested 1 × 1 m and 5 × 5 m plots are sampled. Marking the corners of the 20 × 20 m frame, and sampling nested 1 × 1 m and 5 × 5 m plots within it, establishes the basis for representative repeated measurement of the same fallow plots even if they make a substage transition as they age.

When the 1 × 1 m frame is used for nested subplots, employing several of them within a plot is generally advisable; these should be randomly distributed within the large frame. Use of the species–area curve method described below should be a helpful guide to determining the number of nested subplots required. The China Cluster used five 1 × 1 m nested subplots within each 20 × 20 m plot for various Field Types within the native forest and orchard Land-use Stages. At the Amapá demonstration sites, the Amazonian Cluster used 5 × 5 m nested subplots to characterize understorey regeneration in native forest plots.

House gardens and edges present special sampling problems that prevent the use of the standard frame sizes. Because of the high spatial variability present within most house gardens, we recommend that the entire house garden area be considered the sample plot. Under these circumstances, it is also necessary to measure the area occupied by the house garden. Nested 1 × 1 m and 5 × 5 m subplots may be used within house gardens for herbaceous and other understorey sampling. Edges tend to be linear in shape, and we recommend sampling in 1 m² increments, with the shape and number of increments to be adjusted according to the shape and size of the edges themselves.

Data recording

We recommend two kinds of basic species diversity data: presence and abundance. Recording of presence requires simple listing of species observed in sample plots. Recording of abundance requires additionally listing the number of individuals of each species.

We also recommend collection of data to describe the utility of species surveyed. We suggest that, at a minimum, utility of an individual species be assigned to general categories, such as food, construction, crafts, medicine, commerce and others; this usage follows earlier ethnobotanical literature (e.g. Prance et al., 1987; Pinedo-Vasquez et al., 1990). A single species may be assigned a 'use-value' in as many categories as appropriate. For some purposes, it may be important to distinguish between known 'potential' uses and actual 'intended' uses stated by the farmer of a particular sample area.

Collection of additional data including size and productivity and more detailed ethnobotanical uses of individual plant species may be accomplished at the same time as presence, abundance and utility data are gathered. Detailed information on harvesting and productivity of useful species is generally important. For tree species, diameter and height measurements are also desirable. We encourage observers to maximize the efficiency of field-work by conducting species diversity data collection in concert with other related tasks, as also recommended for agrodiversity data collection in Chapter 5.

Sampling frequency

Some of the variation in plant species diversity present in complex agricultural landscapes is associated with temporal rather than spatial variation. Capturing the temporal component of agrobiodiversity requires repeated sampling of the same plots at appropriate intervals. As a minimum, we recommend a com-bination of seasonal sampling in the first year followed by annual re-sampling in subsequent years as a means of capturing both inter-seasonal and inter-annual change. Farmers should be consulted to determine the timing of re-sampling needed to capture changes in plant species diversity.

Preliminary data analysis

The data collection methods outlined above were designed to permit many kinds of meaningful statistical analyses. Here, we focus on those analyses that can be done by scientists with a minimum of expertise and computational capacity. More detail is presented in Chapter 8. The following sections discuss species richness, analysis of similarity within and between Field Types, species–area curves, and abundance–diversity curves. These constitute the core species diversity analyses and should be reported using a format similar to that illustrated here (Figures 6.1–4).

Species richness and utility

There are four levels at which calculation of species richness is appropriate:

○ within each plot
○ within each Field Type
○ within each Land-use Stage
○ within each demonstration site or survey area.

Species richness within a plot is simply the number of species recorded as present. Within a Field Type, species richness is the number of species present across all of the replicate plots. Within a Land-use Stage, species richness is the number of species present across all of the Field Types that the Land-use Stage contains. And for the landscape area as a whole, species richness is simply the cumulative number of species across all Land-use Stages; in most cases this will be difficult to estimate because it is unlikely that observers will be able to comprehensively sample all Field Types within all Land-use Stages.

At each scale of analysis, the species utility index is simply defined as the percentage of species identified as useful. Figure 6.1 is a sample reporting form for species richness and utility statistics within a priority Field Type.

Species richness and utility sample reporting form

A separate form should be used for each sample area surveyed. Forms for sample areas representing the same Field Type should be submitted together.

(1) DEMONSTRATION SITE: _____

(2) FIELD TYPE: _____   (3) LAND-USE STAGE: _____

(4) Field Type species richness: ___   (4a) Field Type utility index[1]: _____

(4b) Number of 'useful' species[2]:   food _____   construction ____   crafts _____

                                             medicine _____   commerce _____   other _____

(5) Sample area number (and location)[3]: _____

(6) Sample area species richness: _____   (6a) Sample area utility index[1]: ____

(6b) Useful spp.[2]:   food _____   construction ____   crafts _____

                             medicine _____   commerce _____   other _____

(7) Plot number: _____   (8) Plot size: _____ (in metres)

(9) Date sampled: _____   (10) Plot data include[4]: _____

(11) Plot species richness: _____   (11a) Plot utility index[1]: _____

(11b) Useful spp.[2]:   food _____   construction ____   crafts _____

                              medicine _____   commerce _____   other _____

(12) Number and size of nested subplots: _____

(13) Subplot data include[4]: _____   (14) Subplot species richness: _____

(14a) Subplot utility index[1]: _____

(14b) Useful spp.[2]:   food _____   construction ____   crafts _____

                              medicine _____   commerce _____   other _____

[1] (4a), (6a), (11a), and (14a) refer to the percentage of total species identified as 'useful' (see text).

[2] (4b), (6b), (11b), and (14b) refer to the number of individual species identified as 'useful' within each category listed (see text).

[3] Location information for (5) must be retained by the Cluster but need not be submitted.

[4] (10) and (13) refer to the kind of vegetation tallied with the plot or subplot (e.g. woody stems > 1 cm diameter, or herbaceous plants, etc.)

Figure 6.1   *Recommended sample reporting form for species richness and utility*

Similarity analysis

There are a number of similarity indices used in the ecological literature. We selected Sorenson's as the core similarity analysis for species diversity data within and among Field Types and Land-use Stages. We recommend using Sorenson's Similarity Index in three different ways:

○ to compare species composition data taken at different times in the same plot

○ to compare species composition among replicate sample areas within a single Field Type

○ to compare species composition among Field Types within a single Land-use Stage.

The formula for Sorenson's Similarity Index (S_s) is as follows:

$$S_S = 2Tc_{i \cdot j}/(T_i + T_j)$$

where: T_i and T_j = the number of species in sample units i and j, respectively, and $Tc_{i \cdot j}$ = the number of species common to sample units i and j. The result of the formula should be multiplied by 100 and reported as a percentage similarity value. For within-Field-Type analyses, all pairwise comparisons should be made, resulting in a similarity matrix and a mean similarity value (plus or minus a standard error) as illustrated in Figure 6.2. For between-Field-Type analyses, presence data from all replicates should be pooled for each Field Type prior to calculating the index.

(A) Number of species in common among five house gardens at a Macapá demonstration site					
Sample area	1	2	3	4	5
1	37	22	23	19	3
2		38	31	23	2
3			33	21	3
4				31	3
5					4

(B) Sorenson's Similarity Index for all pairwise comparisons among the five house gardens					
Sample area	1	2	3	4	5
1	—	59%	66%	56%	15%
2		—	87%	65%	9%
3			—	66%	16%
4				—	17%
5					—
mean ± standard error = 46 ± 9%					

(A) Matrix illustrating the number of common species in all pairwise comparisons (bold values are total number of species within each sample area.

(B) Matrix illustrating values of Sorenson's Index (S_s) for all pairwise comparisons calculated from data given in (A). Mean and standard error of Sorenson's Index for the set of ten pairwise comparisons are also provided.

Figure 6.2 *Example analysis of species similarity among sample areas within a single Field Type*

Species–area curves

Species–area curves may be constructed for each Field Type by a stepwise calculation of cumulative species richness as data from each replicate plot are added to the total Field Type species richness (Figure 6.3). The two reasons for using species–area curves are to obtain information about the sufficiency of replication within a Field Type, and to compare the species–area relationships of different Field Types, providing they were sampled using the same plot sizes. The species–area curve will 'flatten out' when the number of replicate plots are sufficient to represent within Field Type variation. Differences in the slope and inflection point of species–area curves can reflect differences in both total species richness between Field Types and in the distribution of species richness within them.

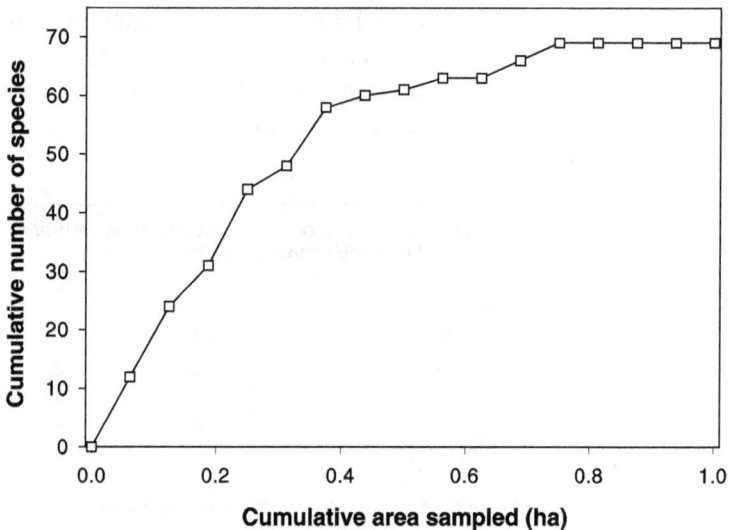

Based on data from sixteen 25 × 25 m plots sampled in an agricultural landscape in Ghana's southern forest-savanna ecotone. Each point represents one plot.

Figure 6.3   *An example of a species–area curve*

Abundance–diversity curves

Abundance–diversity curves are a means of graphically representing the relationship between species evenness and species richness in plots and Field Types. There are several steps involved in producing an abundance–diversity curve; these include:

○ ranking of species by their abundance values
○ calculation of relative abundance values for each species and

○ plotting the relative abundance values against the species ranks (Figure 6.4).

Differences in the slope and shape of the curve reflect differences in species richness and species evenness, and their relationship to one another.

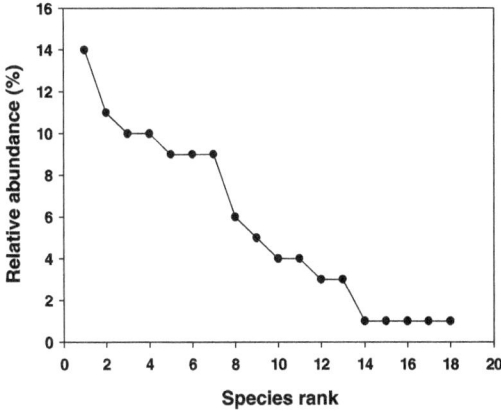

sp. R.	1	2	3	4	5	6	7	8	9	10	11	12	13	14	15	16	17	18	total
Ab.	16	13	12	12	10	10	10	7	6	5	5	3	3	1	1	1	1	1	117
r. Ab(%)	14	11	10	10	9	9	9	6	5	4	4	3	3	1	1	1	1	1	100

Note: sp. R. = species rank (in order of abundance);
            Ab. = abundance of each species;
            r. Ab. = relative abundance of each species (Ab./total abundance x 100)
Based on data from a 20 × 20 m native forest plot in a nature reserve forest at a Yunnan demonstration site (Guo et al., 1998)

Figure 6.4   *An example of an abundance–diversity curve*

Conclusion

This approach to diversity measurement is deeply rooted in biodiversity theory, at least since Whitttaker (1965) suggested the division of diversity into *alpha*, *beta* and *gamma* components. *Alpha* (species) diversity is the diversity within a specific habitat. *Beta* (habitat) diversity is the diversity between habitats. *Gamma* (landscape) diversity is the total diversity of a landscape or region. These guidelines represent a potential application of those basic ecological principles in the context of a quantitative approach to the assessment of diversity in anthropogenic systems subject to long-term management.[1]

7 Household-level agrobiodiversity assessment

GUO HUIJUN, CHRISTINE PADOCH, FU YONGNENG,
DAO ZHILING AND KEVIN COFFEY

Introduction

CHINA IS ONE of the most biologically diverse countries in the world. Scientific estimates have registered 32 500 species of vascular plants, 800 species of freshwater fish, 226 species of amphibians, 313 species of reptiles, 2286 species of birds, and 372 species of mammals (Zhang and Wu, 1996). China is also recognized as a centre of origin of numerous cultivated crops including rice (*Oryza* sp.), peach (*Prunus persica*), soybean (*Glycine max*), tea (*Camellia sinensis*), lychee (*Litchi chinensis*), kiwi fruit (*Actinidia chinensis*), Chinese water chestnut (*Eleocharis tuberosa*), sweet orange (*Citrus sinensis*) and many other species important to agriculture (Wang and Yang, 1995).

Within this country of enormous variety, Yunnan stands out as the most biodiverse of all of China's provinces and is one of the world's biodiversity 'hot spots' for conservation priorities (Myers et al., 2000). In the tropical regions of Yunnan it is estimated that there are 120 species of wild or semi-domesticated relatives of cultivated crops. Studies in the Xishuangbanna Dai Autonomous Region in southern Yunnan have revealed over 1000 species of plants with high economic potential (Wang and Yang, 1995) and 4669 higher plant species (Xishuangbanna Tropical Botanical Garden and Ethnobotanical Department of Kunming Institute of Botany, 1996). Xishuangbanna covers just 0.2 per cent of China's land area, but contains 15 per cent of the country's vascular plant species. The vast richness of genetic resources, both agrobiological and biological, makes conservation in Yunnan a matter of global importance. The preservation of agricultural biodiversity is not just important for the conservation of local livelihoods and land-use systems, but is essential for food and economic security on a global scale.

A history of diversity

The astounding richness of wild biodiversity, semi-cultivated plants, semi-domesticated animals, agroecosystems and local varieties can be linked to the unique geographical and cultural history. Yunnan owes its wealth of ecosystems to its geographical position at the junction of three geological plates and six floristic regions. There are currently 27 separate ethnic minorities, each with a distinctive tradition of resource management. They are the originators, transformers and conservators of much of the biological richness. For example, the Tibetan minority manages the resources of the magnificent snowcapped mountains in the northwest of the province, while the Hani practise swidden fallow agriculture in the tropical southeastern region.

While the geographical and cultural histories have fostered diversity, the political turmoil of the past century has been a major factor in the reduction of biological diversity. Following Liberation, one of the early land policies implemented in rural China in 1949 was the equitable distribution of land among the peasants. By the late 1950s the national government adopted more radical land and resource policies, which shifted the control of agricultural decision making from the household level to an administrative community level. The crucial connection between household decision making and the maintenance of agrodiversity was endangered. Variation in agricultural land-use practices from one household to another, which supports the agricultural innovation that sustains agrodiversity, was jeopardized by policies that promoted uniformity. Control of land tenure and decision making changed to a top-down approach with regional administrators determining land uses. The result was not only a loss of agrodiversity, but also a drastic decrease in efficiency of agricultural production.

In response to the sharp declines in output, the government of China enacted the Household Responsibility System in the late 1970s. This policy restored the household as the basic unit of production and gave farmers an incentive to increase efficiency. The land still remained the property of the state, but a contract gave farmers the right to profit from yields above a specified quota. Though the contract stipulated the crops the farmer must provide for the state, the farmer made all the decisions regarding the maintenance and intensity of management. From the agrodiversity standpoint, considerable variation in farming practices among households soon developed. In time the rules became increasingly lenient as the efficiency of household-level decision making became more and more apparent.

Today, while the influences of past and present policies that threatened agrodiversity are observable, a great variety of land-use systems, technologies, crops and many other factors related to agriculture that were strictly controlled by the state in the recent past, are evident again. Communal labour in agriculture has long existed in China, especially among the minority groups of Yunnan. The changes resulting from the strict collectivization policies of the 1960s and 1970s radically changed traditional community production patterns, and, as mentioned above, decision making was transferred to multi-village-level administrators. The recent return of household decision making has created a unique opportunity to observe how households, previously constrained by uniform production system policies, diverge in new directions, inventing management schemes that both incorporate pre-socialist (traditional) ideas and address the current ecological and socio-economic environment.

Household-level assessment

Yunnan's rich history and diversity have attracted much attention and have been the subject of numerous studies by government and non-governmental research institutes that have contributed toward the conservation of both bio-diversity and agrobiodiversity. Although these studies have made promising advances, the biodiversity of the province, and especially its agrobiodiversity, continues to erode at alarming rates. PLEC made agrodiversity a conserva-tion priority; research and activities carried out in this region include agrobiodiversity assessments (ABA) at the landscape level (for a detailed discussion of ABA see Guo, et al., 1996, 1998). The results of these surveys provided new and innovative findings that could be of use in the formulation of local and global conservation strategies. The following phase of assessment attempted to expand and build on the knowledge acquired from ABA activ-ities, and to further our progressive insights into the field of agrodiversity conservation by developing an assessment method, focused on the house-hold. We have termed the method household-level agrobiodiversity assessment (HH-ABA).

The previous ABA landscape-level analysis studied the diversity within agricultural landscapes at the community level. There was a major effort to involve the local people in the assessment and documentation of the usage and evaluation of plants. The ABA work also attempted to document the complex land-tenure systems in the sample areas. The abundance of existing state-level land-tenure policies, which interact with, and sometimes conflict with, strong traditional regulations, creates a variable and complex land-tenure system that makes its documentation and clear understanding by those from outside the community difficult.

Observation of agrodiversity at the household level is a rare opportunity to study the processes that lead to variation in farming practices, and to enhance agrodiversity throughout Yunnan as well as the rest of the country. The incre-mental divergence of farming activities from the conformity of the past exposes the small changes that gradually lead to agrodiversity. It is extremely rare to find such a transparent view of both the factors promoting agrodiver-sity and the rate at which households become increasingly distinct. Household-level analysis is clearly one of the most useful tools for capturing the dynamics of these changes.

There are five primary goals of the HH-ABA:

○ to promote agricultural systems rich in diversity by documenting and con-serving the plant genetic resources, technologies, and knowledge associated with such systems
○ to explore connections between biodiversity conservation and economic development

○ to document particularly productive and conservationist agricultural technologies and management techniques developed by local households
○ to select the best and most innovative farmers for demonstration activities
○ to observe and promote the exchange of experiences and innovations
among households.

A central objective of the programme is to identify and work with the most
innovative and productive farmers in the community, particularly those who
employ biodiversity to solve problems of production, land management and
pest control. By working at the household level, researchers gain a better
understanding of variation in individual household activities, enabling them
to identify the exceptional farmers and work with them in setting up and
carrying out demonstration activities. Examples of the use of HH-ABA in the
field are presented in Chapters 20 and 21.

An example from Tao Yuan

The story of an expert forester from the Baihualing valley in western Yunnan
provides an excellent example of the type of farmer that PLEC seeks and
HH-ABA identifies. Since 1982 government extension workers have promoted the planting of *Cunninghamia lanceolata*, introduced from elsewhere
in southwestern China, into this valley. The government's purpose was to
encourage afforestation and increase timber resources. The tree needs to
grow for approximately 15 years before it can be sold as a middle-value construction timber.

Today, on the steep slopes of the valley, patches of this species are found
adjacent to some of the most intensive cropping systems in the world. The
maintenance of these complex agricultural systems requires the harmonization of carefully constructed micro-environments for the production of a
tremendous diversity of crops which include maize, rice, coffee, sugar cane,
pumpkin and squash. The innovation and experimentation that created these
remarkable and unique systems has also affected the prescription for timber
production. The focus of the HH-ABA on the household provided an opportunity to improve our understanding of the processes that facilitate
agrodiversity and to locate the farmers responsible for it. The birth of variation takes place not on the community level, but at the individual or
household level, where an innovative smallholder combines an introduced
idea with local knowledge in their living laboratory and, after experimentation, adds a new technology to the intricate web of land uses in the valley.

Community-level analysis of the village of Tao Yuan may have overlooked
the variation from monoculture in the landholdings of one farmer. After
interviewing the farmer and visiting his most prized landholdings, the significance of this became apparent. Along with *C. lanceolata*, two other species,

Phoebe puwenensis and *Toona ciliata*, were dominant in his forest. By adding these two timber species the farmer had increased the value and diversity of his forest. He also staggered the harvesting time of his timber, which encouraged regeneration, increased production, and stabilized yearly earnings.

He collected the seeds of *P. puwenensis* from a nearby nature reserve in Gaoligong Mountain. Through a process of experimentation he learned to dry the seeds and plant them in a semi-shaded area beside a stream close to his house, where he was able to provide a constant supply of water. The seedlings were transplanted into the forest after one or two years. He has had tremendous success and now rents land from his neighbours and manages a total of 3 ha of forest. As well as transplanting the seedlings to his own land, the farmer now sells seedlings to other households in the valley. *P. puwenensis* is a valuable timber species that until now was grown only in small numbers around the house or extracted from an area that is now a nature reserve. This story of an individual benefiting from experimentation and diversification is one of the more striking outputs of HH-ABA.

Variation within communities

Another benefit of household-level analysis is that it offers the opportunity to compare the techniques and technologies employed by the farmers within a community. Community-level analysis must average the variation within a community to get a single set of observations for an entire community. While these observations are important for agrodiversity analysis, they average out the extremes and obscure the variation. It is most often in the extremes that one finds the most interesting and revealing discoveries. Examining unusual farming practices will not only help to locate the innovative farmers, but offers other insights as well. For instance, such observations may help to point out directions in which agricultural systems might develop in the future.

Agrobiodiversity research done over the past several years has investigated and analysed agrodiversity in selected villages. Household-level analysis fitted well into the ongoing research because it integrated the concerns of the biological and social sciences. Of course, such inter-disciplinary approaches also pose problems. PLEC-BAG proposed sampling techniques for the landscape-level analysis that used methods from both the ecological and social sciences. Adjustments were also made in HH-ABA. For instance, ecological approaches to plot size and selection had to take into consideration the current and rapidly changing social reality of landholding in Yunnan villages.

The HH-ABA process begins with the selection of sampling methods. We considered four different approaches to sampling households in a community. An attempt could be made to sample every household in the village. This type of sampling is both time and capital consuming and limits the depth of information that can be collected. Another method is to select only the 'important households'. Households could be selected by speaking with members of the community and asking them to identify the unique or innovative households. This technique is valuable in achieving some of the goals of HH-ABA, but it also has some shortcomings. For instance, unique and strange systems are often the most intriguing but the sampling of purely unorthodox management practices cannot be extrapolated to the whole community. Also, this type of sampling can miss 'hidden' or undervalued details that are important. The opposite of this method attempts to locate and sample the most typical plots in the community according to researchers' direct observations. This method is a quick way to gain a basic idea about the major land-use systems in a village, but it obviously does not meet the needs of HH-ABA because it avoids the exceptions. Finally, random sampling can be used. Predetermined percentages of households within a village can be sampled and then compared and contrasted. This type of analysis is popular in ecological and sociological sampling and is a good way to get an idea of the situation in the village as a whole. However, this still falls short of the goals of HH-ABA because it is possible that the 'best' or 'worst' household may be excluded.

After careful consideration of the benefits and drawbacks of each option, it was decided to adopt more than one sampling method to ensure that all the goals of HH-ABA would be effectively and efficiently met. Random sampling was the basic method used, and additional households were selected by talking to a village leader or specialist. HH-ABA also used preference ranking through short questionnaires. Finally, other households were selected through direct observation by researchers.

HH-ABA is more than counting species

Researchers, government officials and village residents all have important roles to play in the assessments. Local government officials can be helpful in clarifying data or in obtaining information that is not available to villagers. The involvement of government officials, particularly policy makers, is important for promoting agrodiversity-friendly policies. The purpose of the research group is not limited to the recording and analysis of data, but also provides an important link between the village households, government officials and extension agents in the promotion of both agricultural and forestry development as well as conservation. The benefits of continuous discourse among households, researchers, and government functionaries are vital for the success of HH-ABA and of the PLEC initiative as a whole.

Household sampling techniques can be divided into socio-economic and agrobiodiversity assessments. Socio-economic assessment is done using questionnaires, while agrobiodiversity assessments concentrate primarily on the plot inventory. The investigator doing these assessments ensures the direct participation of the owner and manager of the plot, while neighbours or other villagers may provide useful information; the focus of the assessment should be on the primary manager. Both methods should include oral communication with the primary manager, especially for recording management activities. Also, to maintain consistency, one person should be responsible for all research on household activities. Households should be classified according to their land management stages before they are compared.

Figure 7.1 provides a simplified example of one type of output from HH-ABA data analysis. Income/mu (15 mu = 1 ha) and species richness (S/lnA) for eight households from the village of Daka in Xishuangbanna are represented as plots on the graph. Three conclusions can be drawn. First, the graph shows the wide variation in both income/mu ($\sigma = 1.16$) and species richness ($\sigma = 0.77$) between households. Intra-community variation can be a valuable measurement, especially when observed over a period of time. Such analysis is difficult with community-level assessment; however, data from HH-ABA should lead towards the description of variation among households. Second, a correlation can be performed to test the relationship between these two variables among the eight households. The randomness of the distribution suggests no significant correlation ($R^2 = 0.026$) between income and species richness. Finally, while no correlation is shown, it should be noted that some exceptional farmers maintain a high species richness and above-average income. Although expert farmers cannot be selected solely on the basis of two variables, this style of observation can aid in their identification.

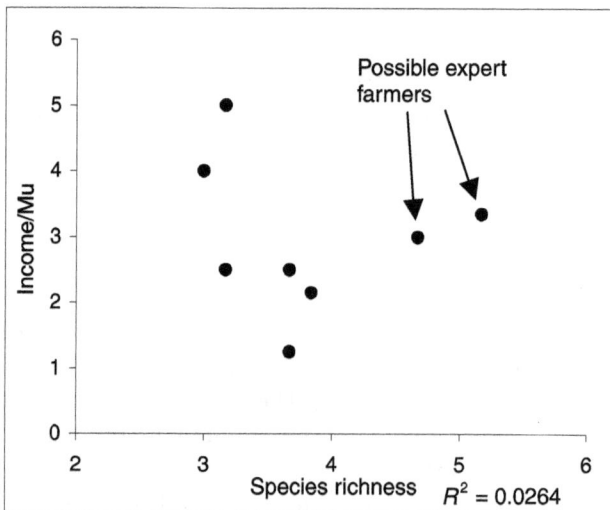

Figure 7.1   *An example of an output from HH-ABA*

Conclusion

Immediately after the implementation of the Household Responsibility System, China's agricultural production systems began to diversify in 900 million directions. It is the great degree of variation among farmers in this short period that makes HH-ABA necessary. This method provides an accurate means of measuring and analysing the process of household diversification during this most fascinating time in China's rich agricultural history, and in the region richest in agrobiological diversity.

8 Quantitative methods for the analysis of agrodiversity[1]

KEVIN COFFEY

Introduction

REPRESENTING SPECIES DIVERSITY in numerical form may at first appear simple. Upon closer inspection, however, the task is somewhat daunting. The ecological literature boasts over 50 diversity indices. This immense list is due in part to redundancy, but each index quantifies a distinct aspect of species diversity. Any of the numerous diversity indices provides a unique numerical value measuring a small and discrete part of the broad and abstract concept of species diversity within an ecosystem. Ideally, an index captures the same aspect of diversity in a collection of samples and provides a viable means for comparison. However, this is not always the case. The ease with which mathematical functions create seemingly exact figures, and the boldness of ecologists in assigning value-laden terms such as diversity and richness to these figures, often leads to misuse or misunderstanding. Measurements and calculations inherently contain assumptions that must be examined before any conclusions can be drawn.

Beyond the various assumptions associated with each index, there is often confusion over what the value actually means. In the complex agroecosystems measured within demonstration sites, one must be careful to avoid over-generalizing the meaning of these measurements. For example, species diversity index calculations are based upon a tally of species within a particular plot. Conclusions drawn from such calculations can go no further than to suggest that one plot has a higher species diversity than another. A sample with a higher diversity index is not inherently superior, nor can any conclusions on genetic diversity, soil quality, productivity, or the general health of a system be assumed from this measure.

This short summary provides a concise (and therefore incomplete) overview of indices, measurements and methods of agrodiversity analysis. The chapter is divided into two sections. The first section deals specifically with measurements of species diversity. For more information on this topic, please consult Magurran (1988) and Krebs (1998). The second section presents a sample of statistical methods that can be used to test agrodiversity-based hypotheses. References for each technique are cited in the text.

Diversity indices

The diversity measurements discussed below quantify species diversity within a group of plots or sample areas for means of comparison. While numerous formulas can be used for such comparisons, most of them, with the exception of β diversity, are based upon three concepts: species richness, heterogeneity, and evenness.

Species diversity

The simplest measure of species diversity is *species richness*, a count of the number of species in a particular sample. Ecologists have moved beyond this simple calculation to include more information about the sample in the index. The most common among these are a group of diversity indices that divide the number of species by a measure of abundance. The nature of PLEC agrobiodiversity inventories accommodates the use of the total number of individuals as abundance. This does not always have to be the case. Other forms of abundance can be used, such as land cover, biomass or yield (Garcia, 1992). In the context of agroecosystems, this point is an important one. When abundance is measured by the number of individuals, factors like size are extremely important. A plot containing one tree and 50 rice plants would have very different diversity values if abundance were measured in land cover compared with abundance measured in number of individuals. For this reason, land-use systems that tend to have species close to the same size or samples with qualifiers (e.g. diameter at breast height >5 cm) are most accurately represented by the following indices:

Margalef : $\qquad\qquad\qquad D_{mg} = (S - 1)/\ln N$
Menhinink: $\qquad\qquad\qquad D_{mn} = S/\sqrt{N}$
Gleason: $\qquad\qquad\qquad G = S/\ln N$

where S = the number of species and N = the number of individuals.

Worked example

The truncated data sets, with species names and abundances of trees in fallows of similar age owned by different farmers in Amazonian demonstration sites, presented in Table 8.1 will be used for the calculation of species diversity indices.[2]

Table 8.1   Data set of species names and abundance of trees in fallows in Amazonia

	Sample A			Sample B	
	Species	*Abundance*		*Species*	*Abundance*
1	Eugenia cumini	2	1	Rollinia mucosa	6
2	Symphonia globulifera	4	2	Protium pubescens	3
3	Carapa guianensis	5	3	Theobroma cacao	7
4	Carapa sp.	2	4	Herrania mariae	5
5	Banara guianensis	3	5	Dendrobangia boliviana	6
6	Stryphnodendron guianense	10	6	Ficus pertusa	2
7	Licania macrophylla	5	7	Rheedia macrophylla	5
8	Ficus pertusa	2	8	Symphonia globulifera	4
9	Rheedia macrophylla	4			
10	Musa spp.	2			

The Margalef Index can be calculated based upon the summary Table 8.2:

Table 8.2   Number of species and individuals in the two samples

	Sample A	Sample B
Species richness (S)	10	8
Individuals (N)	39	38

$$\text{Margalef Index} = S - 1/\ln(N)$$
$$\text{Sample A } D_{mg} = 10 - 1/\ln(39) = 2.46$$
$$\text{Sample B } D_{mg} = 8 - 1/\ln(38) = 1.49$$

As the data indicate, sample A has a higher species diversity than sample B and both samples contain a similar number of individuals, therefore sample A has a higher D_{mg} than sample B. Mehinink and Gleason indices show similar results (D_{mn}: A = 1.6, B = 1.3; G: A = 2.7, B = 2.2).

Heterogeneity

Heterogeneity indices quantify both species richness and evenness in a single measure. Evenness refers to the distribution of abundances in a sample. The two data sets provided above show different levels of evenness. A quick glance at the samples reveals that the species abundances in sample A are less evenly distributed than those in sample B. Comparing both data sets using heterogeneity indices would make the samples appear more similar in diversity than the species diversity indices calculated above. This is because sample A has a higher species richness than sample B, but sample B is more evenly distributed than sample A.

The three most common heterogeneity indices are:

Shannon[3] (based on information theory):
$$H' = -\Sigma p_i * \ln p_i$$

where p_i equals the relative abundance of the ith species = (n_i/N).

Simpson (based on probability):
$$D = \Sigma(n_i(n_i - 1))/N(N - 1)$$

where n_i equals the number of individuals in the ith species and N equals the total number of individuals.

Brillouin: $$HB = (\ln N! - \Sigma\ln(n_i!))/N$$

where n_i equals the number of individuals in the ith species and N equals the total number of individuals.

Worked example:

Tables 8.3 and 8.4 show the steps used to calculate the Shannon Index. In column three, the relative abundance (p_i) is calculated for each species. The natural log of the relative abundance is calculated in the next column. In the last column, the two previous columns are multiplied. Finally, all values in the last column are added. The number is then changed from negative to positive.

Table 8.3   Shannon Index calculation for sample A

Species	Abundance	p_i	$\ln(p_i)$	$p_i * \ln(p_i)$
			Sample A	
Eugenia cumini	2	0.051	− 2.970	− 0.152
Symphonia globulifera	4	0.103	− 2.277	− 0.234
Carapa guianensis	5	0.128	− 2.054	− 0.263
Carapa sp.	2	0.051	− 2.970	− 0.152
Banara guianensis	3	0.077	− 2.565	− 0.197
Stryphnodendron guianense	10	0.256	− 1.361	− 0.349
Licania macrophylla	5	0.128	− 2.054	− 0.263
Ficus pertusa	2	0.051	− 2.970	− 0.152
Rheedia macrophylla	4	0.103	− 2.277	− 0.234
Musa spp.	2	0.051	− 2.970	− 0.152
			Sum	− 2.149

Table 8.4   Shannon Index calculation for sample B

Species	Abundance	p_i	$\ln(p_i)$	$p_i * \ln(p_i)$
			Sample B	
Rollinia mucosa	6	0.176	− 1.735	− 0.306
Protium pubescens	3	0.088	− 2.428	− 0.214
Theobroma	7	0.206	− 1.580	− 0.325
Herrania mariae	5	0.147	− 1.917	− 0.282
Dendrobangia boliviana	6	0.176	− 1.735	− 0.306
Ficus pertusa	2	0.059	− 2.833	− 0.167
Rheedia macrophylla	5	0.147	− 1.917	− 0.282
Symphonia globulifera	4	0.118	− 2.140	− 0.252
			Sum	− 2.134

The summary Table 8.5 highlights the difference between species richness and heterogeneity indices. The species diversity index (D_{mg}) shows a much greater difference between the two samples than does the heterogeneity index. The choice of index reflects the different aspect of diversity one wishes to measure. In this case, either species richness and/or evenness.[4]

Table 8.5   Comparison of indices

	Sample A	Sample B
Species richness (S)	10	8
Individuals (N)	39	38
Margelef (D_{mg})	2.46	1.49
Shannon (H')	2.15	2.13

Rank abundance

One of the most effective methods for representing both species richness and evenness of a sample is a rank abundance graph. This technique is rarely used when a large number of samples are being compared because it does not provide a single measure.[5] To create the graph, both samples are ranked from highest abundance to lowest (Table 8.6).

Table 8.6   Species abundance rank

	Sample A			Sample B	
	Species	Abundance		Species	Abundance
1	Stryphnodendron guianense	10	1	Theobroma cacao	7
2	Carapa guianensis	5	2	Rollinia mucosa	6
3	Licania macrophylla	5	3	Dendrobangia boliviana	6
4	Symphonia globulifera	4	4	Herrania mariae	5
5	Rheedia macrophylla	4	5	Rheedia macrophylla	5
6	Banara guianesis	3	6	Symphonia globulifera	4
7	Eugenia cumini	2	7	Protium pubescens	3
8	Carapa sp.	2	8	Ficus pertusa	2
9	Ficus pertusa	2			
10	Musa spp.	2			

The rank is plotted on the *x*-axis and abundance on the *y*-axis. The graph visually represents the evenness (represented by slope) and species richness (represented by length) of both samples. Figure 8.1 covers every aspect of diversity quantified in the previously mentioned indices.

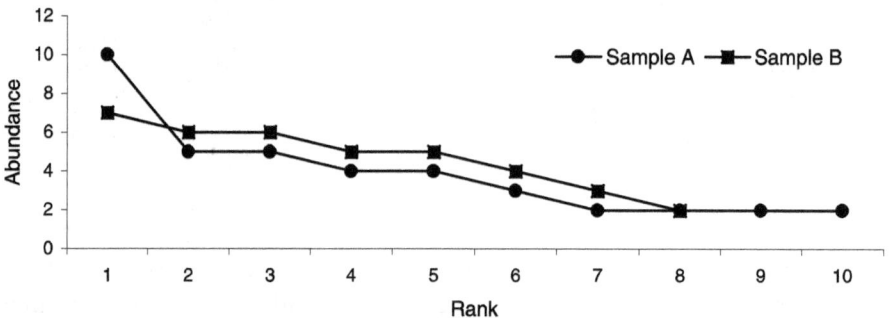

Figure 8.1   *Rank abundance of two samples*

Tools for agrodiversity analysis

After one or more measurements for agrobiodiversity have been chosen and calculated for each sample area, the next step is to compare the results. The literature has provided examples of the use of these indices (Pinedo-Vasquez et al., 2000, also Chapter 15; Coffey, 2000; Guo et al., 2000b, also Chapter 7). The biodiversity and agroecosystem literature also contains many examples of this type of analysis (Garcia, 1992; Baudry, 1993). Conclusions drawn from the analysis of these indices most often rely on a set of variables that fall under the category of agrodiversity data. This section summarizes a few of the more popular methods of agricultural diversity analysis.

Agrodiversity matrix

One such method of analysis, recommended by Pinedo-Vasquez (personal communication, 2001), is an agrodiversity matrix. The goal of the matrix is to reveal the changes in agrobiodiversity (in the form of an index) under a set of agrodiversity variables. Before a matrix can be created, the set of agrodiversity variables to be included in the matrix must be selected. Table 8.7 lists an example of variables used by the Amazonia Cluster. These variables are based on direct observations by the Amazonia Cluster and recommended in the literature (specifically Brookfield et al., 1999, also Chapter 5).

Table 8.7 Agrodiversity variables

Management	Farmer information	Biophysical
Selective weeding	Household size	Soil type
Slashing	Age	Field Type
Burning	Tenancy	Land form
Ploughing	Off-farm employment	Area
Tilling	Labour supply	Slope
Mounding	Transport	Aspect
Ridging	Access to loans	Elevation
Organic fertilizer	Wealth category	
Inorganic fertilizer	Market access	
Planting material		
Thinning		
Weeding		

Using variables from Table 8.7, a matrix can be created, displaying changes in agrobiodiversity measurements in relation to the presence, absence or degree of the variables. Table 8.8 is an example using management variables. The results of *t*-tests based on species richness and the Shannon Index are also included. The *t*-tests test for a significant difference between samples grouped by the presence or absence of a single variable.

β diversity

β diversity is a concept unlike the previous indices because it identifies the differences in species composition between plots or subplots within a single sample or group of samples. The quantification of β diversity can be described as a measurement of species turnover among plots or samples. The measurements provide an estimate for how different a group of samples are. One measure of ß diversity is Whittaker's:

Table 8.8 Agrodiversity matrix

Variables	Sample size	Avg species richness	Individuals per m^2	Avg Shannon	Avg D_{mg}	Species richness t-test			Shannon Index t-test		
						t	DF	Sign.	t	DF	Sign.
Selective weeding	22	23.5	6.1	2.97	2.17	−1.42	33	0.164	−9.68	33	0.000
No selective weeding	13	25.6	3.4	3.61	3.75						
Slashing	16	15.1	0.4	2.21	4.76	−5.60	33	0.000	−6.34	33	0.003
No slashing	19	29.1	1.3	3.76	6.00						
Burning	14	26.4	4.5	3.24	5.67	−3.73	33	0.001	−3.24	33	0.001
No burning	21	12.2	5.4	1.56	4.55						
Ploughing	13	17.8	4.0	2.93	5.69	−4.20	33	0.003	−5.26	33	0.007
No ploughing	22	27.9	5.6	3.99	5.41						
Tilling	18	19.8	4.6	2.12	4.11	−1.72	33	0.107	−4.56	33	0.053
No tilling	17	24.4	1.8	2.65	5.70						
Mounding	19	25.7	0.9	3.23	5.51	2.50	33	0.018	−1.45	33	0.016
No mounding	16	17.6	4.8	2.13	4.72						
Organic fertilizer	14	23.6	6.5	2.14	4.87	−3.34	33	0.012	−2.32	33	0.009
No organic fertilizer	21	28.2	0.4	3.11	5.75						
Inorganic fertilizer	15	22.9	5.7	2.35	4.03	−0.22	33	0.824	−1.67	33	0.624
No inorganic fertilizer	20	22.3	1.9	2.53	4.45						
Thinning > 3 days	7	12.4	4.8	1.34	4.09	−7.38	33	0.000	−5.52	33	0.000
Thinning 1–2 days	15	22.0	0.8	2.34	4.34						
No thinning	13	25.9	1.6	2.65	5.93						
Weeding > 3 days	8	23.4	4.6	2.65	5.27	−6.24	33	0.000	−4.32	33	0.000
Weeding 1–2 days	16	25.5	5.6	3.12	4.49						

Whittaker: $$\beta = S/\alpha - 1$$

where S = the total number of species in the sample and α = the average number of species in each sub-sample.

Worked example
This measure includes only the number of species found in each sub-sample and the total number of species. The abundance of each species does not influence the outcome. Table 8.9 shows the number of species found in eight plots (10 × 10 m) inventoried in two samples of Amazonia fallow.

Table 8.9 Species abundance

Plot	Number of species recorded	
	Sample A	*Sample B*
1	3	1
2	5	2
3	6	4
4	4	3
5	7	5
6	3	1
7	5	4
8	8	4
α	5	3

The S values (from Table 8.2) for sample A and sample B are 10 and 8, respectively. α is calculated by averaging the number of species found in each plot for each sample (Table 8.9).

The calculation would be as follows:

$$\text{Sample A: } \beta w = S/\alpha - 1 = 10/5 - 1 = 2 \quad -1 = 1$$
$$\text{Sample B: } \beta w = S/\alpha - 1 = \quad 8/3 - 1 = 2.7 - 1 = 1.7$$

The higher value of βw for sample B shows that the plots within sample B have greater differences in species composition than the plots in sample A. Table 8.10 summarizes the five values calculated for the two fallow samples.

Table 8.10 Comparison of indices

	Sample A	*Sample B*
Species richness (S)	10	8
Individuals (N)	39	38
Margelef (D_{mg})	2.46	1.49
Shannon (H')	2.15	2.13
Whittaker (βw)	1	1.7

Measurements of similarity

Measures of similarity are almost as numerous as measures of species diversity. The purpose of these functions is to quantify the similarity between two or more samples. The two measurements of similarity explained below are called binary similarity coefficients, as opposed to quantitative coefficients, because they rely on presence/absence data (in this case species). The first step in calculating binary coefficients is to create a table in the form of Table 8.11.

Table 8.11 **Binary similarity matrix**

		Sample B	
		Number of species present	Number of species absent
Sample A	Number of species present	a	c
	Number of species absent	b	–

The internal boxes have been labelled a, b, and c:

where a = number of species in both sample A and sample B, b = number of species in sample B and not in sample A, and c = number of species in sample A and not in sample B.

The following binary similarity coefficients can be calculated using the information above:

Jaccard: $S_j = a/(a + b + c)$

Sorensen: $S_s = 2a/(2a + b + c)$

The two coefficients are similar. The only difference is that Sorensen gives more weight to the species that are present in both communities, and therefore less weight to species that are present in only one community. The choice of measurement should depend on whether the absence of a species is considered as important as the presence of a species. The similarity measurement can be calculated for a group of two or more samples and results can be assembled in a matrix of similarity. The example below explains this procedure.

Worked example

Seven truncated inventories of house gardens (labelled A–G) from the village of Mazagão in the Brazilian state of Amapá are used in this example (Table 8.12).

Table 8.12 **Inventory of house gardens**

Species list	House gardens						
	A	B	C	D	E	F	G
Calophyllum brasiliense	x			x			x
Campsiandra laurifolia	x		x	x			
Ficus maxima	x		x				x
Ficus pertusa	x						x
Gustavia augusta	x	x	x	x		x	x
Hura crepitans	x				x		
Licaria mahuba			x	x	x		
Mora paraensis		x				x	x
Olmedia caloneura				x			
Protium pubescens	x	x	x				x
Sapium lancelatum			x	x			x
Virola surinamensis	x		x	x	x	x	

Table 8.13 was created using the model of Table 8.11.

Table 8.13   Species in house gardens

		House garden B	
		Number of species present	Number of species absent
House garden A	Number of species present	2	1
	Number of species absent	6	–

The Jaccard calculation would be as follows:

$$S_j = a/(a + b + c) = 2/(2 + 6 + 1) = 2/9 = \mathbf{0.22}$$

The procedure is repeated until every sample has been compared with all the other samples. A matrix is then created to compare similarity measurements (Table 8.14). One half of the table is left blank because the similarity of 'A to B' = 'B to A' and so on.

Table 8.14   Jaccard similarity matrix

	A	B	C	D	E	F	G
A	–	–	–	–	–	–	–
B	0.22	–	–	–	–	–	–
C	0.50	0.25	–	–	–	–	–
D	0.36	0.11	0.56	–	–	–	–
E	0.22	0.00	0.25	0.25	–	–	–
F	0.22	0.50	0.25	0.25	0.20	–	–
G	0.56	0.43	0.40	0.27	0.30	0.25	–

The Jaccard similarity matrix shows the similarity in species composition for all seven house garden samples. With large data sets, the use of this type of matrix becomes cumbersome. The results of the matrix can be represented in a more useful format by incorporating a multivariate analysis technique called cluster analysis.

Cluster analysis[6]

Cluster analysis is used to group a large set of samples into one or more clusters. The previous example provides a matrix of similarity that can serve as a basis for grouping the samples according to their similarity to each other. It is important to remember that the Jaccard matrix compares only the species inventory of each sample. The grouping of clusters according to this matrix will simply provide a new way of looking at the presence/absence of species among samples. Any of the numerous formulae for similarity (including binary and quantitative) can be used for cluster analysis. This example uses Jaccard because the values have already been calculated (see Table 8.14).

Cluster analysis often requires multiple steps of calculation and recalculation and is most often performed using a computer. For this example, the simplest method of cluster analysis, called single linkage clustering, is used. With the help of a computer, average linkage clustering can be more accurate (Krebs, 1998: 397). The example below outlines the steps for single linkage clustering.

Worked example
The first step in single linkage clustering is to find the samples with the highest similarity. The Jaccard similarity matrix shows that samples A and G, as well as samples C and D have an $S_j = 0.56$. These two groups will form the first two clusters, AG and CD.

The second step is to look for the samples with the next highest similarity value. Samples A and C, as well as samples B and F have similarity measures of 0.50. At this point, samples B and F are grouped together as a cluster. Samples A and C are already members of separate clusters. In single linkage clustering, the clusters assume the highest similarity measures of the samples within it.[7] Therefore clusters AG and CD become one cluster called AGCD.

The procedure is repeated until all of the clusters are joined together. The tree diagram (Figure 8.2) shows the results of the cluster analysis based on the Jaccard similarity matrix.

The tree diagram can be used to assess whether an agrodiversity variable contributes to similarities in species composition. It is important to remem-

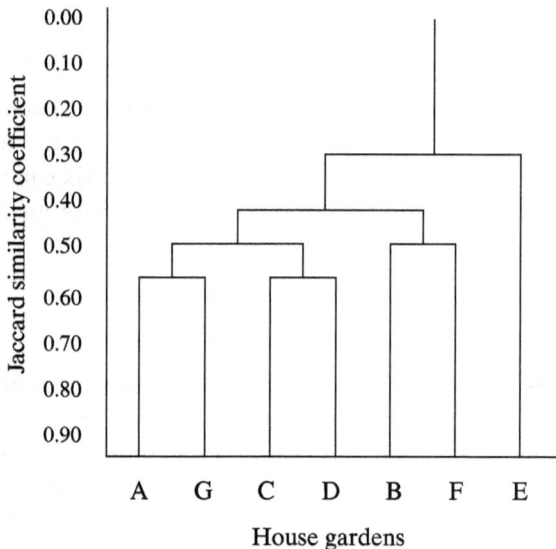

Figure 8.2   *Cluster analysis of house gardens*

ber that the diagram represents only the analysis of species lists for similarity. The interesting analysis comes from the inclusion of the many other data associated with each sample (Garcia, 1992; Baudry, 1993; Pinedo-Vasquez et al., 2000). At similarity 0.50, the samples can be grouped into three clusters AGCD, BF, and E. Table 8.15 compares a few attributes of each cluster.

Table 8.15 Comparison of attributes of the three clusters

	Samples	Average area (ha)	Average Shannon Index	Average annual income	Average household size
Cluster 1	AGCD	6.4	2.56	2495	5.6
Cluster 2	BF	3	1.10	1725	4.5
Cluster 3	E	10	4.35	5000	6

Principal component analysis

Principal component analysis (PCA) and cluster analysis are both statistical tools for condensing data sets. While cluster analysis reduces the number of samples, PCA reduces the number of variables. PCA can be especially helpful in analysing agrodiversity data because these data tend to be collected on a large set of management and environmental variables. Many of these variables will be correlated. PCA converts a large set of correlated variables into a smaller set of uncorrelated variables. This smaller set of uncorrelated variables represents most of the variation within the larger set, but the correlated variables are hybridized.

For example, take a survey of 500 car owners asked the following three questions: (1) How many miles do you drive a year? (2) What is your income and (3) How much did you pay for your car? Where cluster analysis of the results of this survey would group the people (samples) in the survey according their responses to the questions, PCA groups the questions (variables). There would probably be a strong correlation between the cost of a person's car and their income. By grouping the correlated variables, the dimensionality of the data can be reduced to two uncorrelated variables. One variable would represent a car cost/income hybrid and the other variable might represent miles driven per year. With only three variables this type of analysis may appear like a simple and perhaps useless statistical tool. For data sets with numerous variables that are correlated in less obvious ways, however, this technique can be an extremely effective procedure to simplify the data set and reveal interesting trends while maintaining most of the information in the original data set.

A statistical software package is necessary to perform PCA. There are many packages available, including SPSS, MiniTab, and S-Plus. The programs speed through the elaborate calculations with a few clicks of the mouse. This short introduction will touch on a few of the concepts behind PCA and techniques for analysing the outputs. As recommended with the indices and methods mentioned previously, it is useful to consult one or more of the many statistics books that cover this topic in detail. Dunteman (1989) and Jolliffe (1986) are useful PCA references.

The basic concept of PCA is best described graphically. Figure 8.3 contains a set of highly correlated data points plotted with variable A on the *x*-axis and variable B on the *y*-axis. The goal of PCA for this data set would be to reduce the number of variables from two to one. A principal component is essentially a line drawn through the data points that accounts for as much variation within the data as possible. This would be the line labelled PC 1. This idea is similar to, but different from, a regression. The sum of the distances between all of the data points and PC 1 is smaller than any other line that can be drawn on the graph. This line is said to have the highest variance. This is the first principal component. Simply put, if one were to use a vice to flatten all of the data points on to one line, PC 1 would require the least amount of compression. PC 1 would therefore retain the most information about the original data.

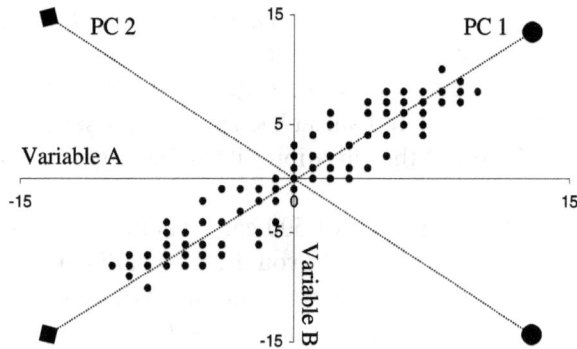

Figure 8.3   *Graphical representation of PCA*

The results of this 'flattening' of data points on to PC 1 are shown below. The shapes at the end of the line have been added only as reference points and would not normally appear on this type of graph. The data on the PC 1 number line represent a hybrid of variables A and B. The original data set can be projected on to PC 1 in order to convert two variables (two dimensions) into one principal component (one dimension). The relationships between

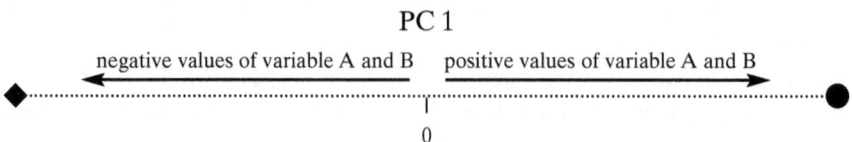

Figure 8.4   *Graphical representation of principal component 1 (PCI)*

the principal component (PC 1) and the original variables are labelled on the number line.

The data in this example are highly correlated, therefore the first principal component is an excellent representative of the original data. A second principal component (PC 2) can be drawn to account for the variance that the first principal component missed. This line is drawn perpendicular to PC 1. PC 2 represents a much smaller portion of the variance in the original data. A number line of PC 2 is labelled below:

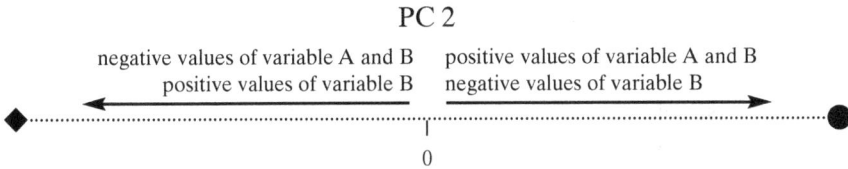

PC 2

| negative values of variable A and B | positive values of variable A and B |
| positive values of variable B | negative values of variable B |

0

Figure 8.5   *Graphical representation of principal component 2*

In PCA there will always be as many principal components as there are original variables. In this case there were two original variables and two principal components. The purpose of PCA is to reduce the number of variables. By calculating the variance represented by each principal component in the original data set, less significant principal components can be eliminated. Selecting principal components based on their variance will be discussed in more detail in the next example. Without calculating the variance, it is obvious that PC 1 closely follows the trend in this data set. For future analysis PC 1 can replace variables A and B. This reduces the dimensionality of the data set from two to one, while maintaining most of the information in the original data set.

In practice, PCA is performed on data sets with more than two variables. To include another variable in this example, a third axis can be added and each of the data points could be plotted in three-dimensional space. If the third variable was strongly correlated with the other variables, the cluster of points might take a cylindrical shape like a pen. The first principal component would then run lengthwise through the centre of the cylinder, like the ink cartridge, to capture the maximum variance in the data set. Two other principal components would be drawn perpendicular to the first component. Because a pen shape represents highly correlated data, the first principal component would be representative of all three variables. These three correlated variables would be reduced to one principal component.

It is impossible to imagine what a fourth or fifth variable added to this analysis would look like graphically, but it is statistically possible to calculate principal components in more than three dimensions. This is where PCA becomes a powerful analytical tool. The purpose of this section was to explain the concepts behind reducing the dimensionality of data. The graphs shown above are not outputs in standard PCA. Statistical programs calculate the functions and their variance automatically. The following example explains the principal component analysis of five variables. It uses the statistical program SPSS to calculate the principal components and their variance.

Worked example

Table 8.16 displays the number of man-days per year spent performing five management activities in ten fallows owned by ten different farmers. The data are not real. Notice the first six rows follow a similar pattern (each column is roughly half of the previous column, while the broadcasting and transplanting columns are roughly equal). The last four rows follow a different pattern (the last three columns are roughly equal and the second column is roughly half of the first column). These trends have been purposely added to the data to show how PCA can help to find such trends in larger and less correlated data sets.

Table 8.16   Number of man-days per year spent in management tasks in ten fallows

Farmer	Weeding	Thinning	Pruning	Broadcasting	Transplanting
1	32	16	8	4	4
2	46	23	11	5	6
3	52	26	13	7	6
4	25	13	7	3	4
5	30	15	7	3	3
6	20	10	5	2	3
7	34	17	9	9	8
8	36	18	9	10	9
9	35	16	8	8	9
10	31	15	7	8	8

Tables 8.17 and 8.18 show the results of the PCA. These tables show a common output format for statistical packages. Table 8.17 is basically a summary of all the principal components. Since there were originally five variables there are five principal components. The important statistic to look for in this table is the variance that each of the components represents. The first column lists the components. The second column contains the total variance, also called eigenvalues or latent roots, for each component.[8] The next column lists the percentage variance for each component. The percentage variance is calculated by dividing the total eigenvalue for each component by the sum of eigenvalues for all five components. This is an important value because the percentage variance shows the degree to which a particular component represents the original data set. The final column is a running sum of the variances in the previous column. This cumulative column is a meaningful measure because it is used to eliminate inconsequential components.

The purpose of PCA is to reduce the number of variables. Since the number of components will always equal the number of variables, there needs to be a

Table 8.17   Output of principal component analysis (PCA)

Component	Total variance	% of variance	Cumulative % variance
1	3.561	71.218	71.218
2	1.375	27.491	98.709
3	4.461E-02	0.892	99.601
4	1.788E-02	0.358	99.959
5	2.049E-03	4.097E-02	100.000

selection process to extract the top few components that represent the heart of the data. There is more than one way to decide how many components to keep, but the simplest way is to maintain around 80 per cent of the variance. This example purposely uses a highly correlated data set. The first component containing 71.2 per cent of the variance is unusually high. It is also rare to have the first two components (of five) representing 98.7 per cent of the data. This means that the first two components maintain more than enough of the variance. Normally the group of principal components would have a cumulative variance closer to 80 per cent. The last three components represent such a small part of the variance that they are irrelevant.

Details about the first two principal components are listed in Table 8.18. The latent roots of each variable are listed for both components. Together, the latent roots supply a formula for the principal component. The degree and sign of the latent roots reveal information about the relationship between the variables for that component. The part of the table labelled 'Commonalties' is important because it shows the degree to which the initial data are represented in the two components (extracted) for each of the five variables. In this example the variances are all very high, but if the extracted value was drastically lower for one or two of the variables, it may be a good idea to exclude it from the analysis.

Table 8.18   Component matrix

	Component			Commonalities	
	1	2		Initial	Extraction
Weeding	0.953	-0.318	Weeding	1.000	0.990
Thinning	0.925	-0.378	Thinning	1.000	0.998
Pruning	0.928	-0.353	Pruning	1.000	0.986
Broadcast	0.722	0.677	Broadcast	1.000	0.980
Transplant	0.659	0.740	Transplant	1.000	0.982

Note that all of the latent roots for the first principal component are positive. The numbers are all relatively close. Weeding, thinning, and pruning are weighted almost equally (0.953, 0.925 and 0.928) and broadcast and transplanting have slightly lower, but similar weights (0.722 and 0.659). It can be concluded that this first component represents a positive relationship between all five variables. Remember that the first six columns in the original data set were purposely created to exhibit this relationship. The second component represents the last four rows of Table 8.16. These rows have higher broadcasting and transplanting numbers and slightly lower than average values for the other management techniques. This is displayed in the second principal component, with negative values for the first three variables and positive values for broadcasting and transplanting. Through this analysis, the information contained in the five original variables has been reduced to only two principal components. A graph can now be created using the formulas from the principal components. The formula for PC 1 would be:

$$(\text{weeding} \times 0.953) + (\text{thinning} \times 0.925) + (\text{pruning} \times 0.928) + (\text{broadcast} \times 0.772) + (\text{transplant} \times 0.659)$$

Therefore, PC 1 for the first fallow would be:

$$32(0.953) + 16(0.925) + 8(0.928) + 4(0.772) + 4(0.659) = 57.924$$

Table 8.19 shows the values for the two new principal components calculated using the weights from the Table 8.18.

Table 8.19 Principal component analysis of management in fallows

Farmer	Weeding	Thinning	Pruning	Broadcasting	Transplanting	PC1	PC2
1	32	16	8	4	4	57.924	−13.38
2	46	23	11	5	6	82.425	−19.38
3	52	26	13	7	6	94.158	−21.774
4	25	13	7	3	4	46.898	−10.344
5	30	15	7	3	3	52.804	−13.43
6	20	10	5	2	3	36.171	−8.331
7	34	17	9	9	8	67.909	−8.402
8	36	18	9	10	9	72.101	−7.999
9	35	16	8	8	9	66.936	−7.926
10	31	15	7	8	8	60.652	−6.663

The ten fallows are plotted on the graph in Figure 8.6. Through PCA the five original variables have been reduced to two variables on the x and y axes, PC 1 and PC 2, respectively. The first six fallows are plotted as circles and fallows 7–10 are square plots. Notice the linear relationship between the circle plots. Remember that the data for those fallows were created by starting with an arbitrary number and roughly halving it for each variable. The graph captures that relationship. Fallows 7–10 form a separate cluster. From here the new principal components can be used as variables in subsequent analysis. This example is quite simple, with five variables and only ten samples, but PCA can be used on much larger sets of data to reveal hidden patterns within data sets that have been obscured by a profusion of variables.

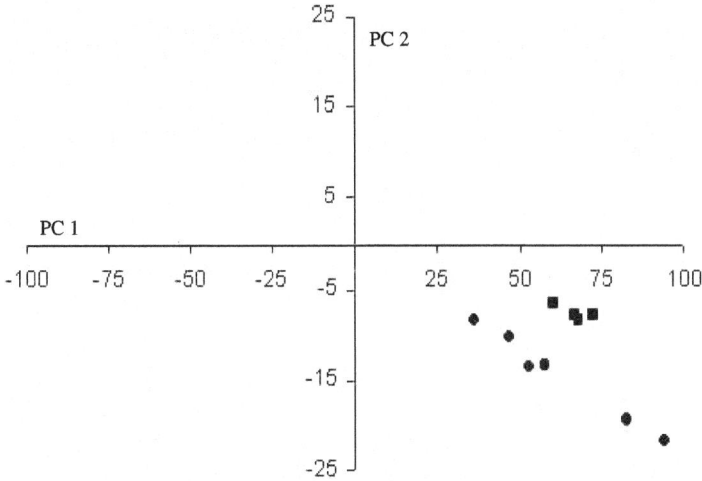

Figure 8.6   *Graphical representation of PCA of ten fallows*

Conclusion

An endless series of variables influence agrodiversity. In collecting and analysing agrodiversity data the researcher is constantly faced with the same dilemma. Which variables are important? The formulas and analysis techniques discussed above provide a few approaches to this question.

Part II. Focus on people

9 Spotting expertise in a diverse and dynamic landscape

CHRISTINE PADOCH

DIVERSITY IS CENTRAL to the PLEC project, and the conservation of biological diversity in agricultural landscapes is an essential goal. The project strives to achieve that goal through the understanding and promotion of agrodiversity, a complex notion that integrates biological diversity at various levels, diversity of agricultural technologies, and diversity in the organization of agricultural enterprises. This task was undertaken because there is considerable evidence that maintenance of agrodiversity is often of benefit to small farmers, their communities and their families.

Most of our participants, like most farmers, are well aware that not all diversity – biological or otherwise – leads to an improvement in the lot of small farming households. Specialization has often been a road to success. But the road of ever-increasing simplicity has been well documented, tested, and promoted by agronomists, foresters, and the government and private organizations that they serve. In recent decades the path of maintaining or increasing diversity in agricultural production has been given little official attention, less encouragement, and far less funding by research and agricultural development institutions. Diversity in production, nevertheless, has been the road chosen by innumerable small farmers the world over. We have tried to work with this large constituency.

Not only farm products and farming systems, but farmers themselves are diverse. Within even the smallest, most 'traditional' community, farmers differ in their capabilities, their perceptions, their preferences, and their expertise, as well as in their goals. This diversity yields both the project's greatest asset and the biggest problem for its participants to solve. In the multitude of possibilities, it has not been easy to decide on which activities our participants should spend their limited time, funds and energies. What should we be promoting and why? In a project committed in the broadest sense to furthering actions that lead to an enrichment in farmers' livelihoods in biologically diverse environments, how do we identify those technologies, patterns and farmers, that should receive backing?

Having rejected the guidance of important, but often simplistically interpreted, ideals of promoting the most 'modern', 'efficient' or the 'highest yielding' that drive most traditional agricultural research and extension projects, PLEC workers have had to resist the temptation to substitute equally simplistic ideals of advancing systems that are the most biologically diverse, the 'most traditional' or the 'most indigenous'. The successful, diversity-promoting systems that we encourage are usually those that are flexible and which can be shown to have evolved in ever-changing physical, biological, political and economic environments. Contrary to common assumptions about biodiversity conservation projects, PLEC does not look with disfavour on new technologies. We seek to foster ways of farming and managing resources that make use of locally developed knowledge, expertise and practice, rather than promoting, as do most development projects, only new technologies that reject or ignore local practice. Among the most promising of project activities are those that help to advance systems that are distinctly hybrid in nature – those management patterns that take the insights of local practice and integrate them with the most modern techniques, to come up with wholly new, distinct, but locally well-adapted, resource-use patterns.

Experts and expertise in a PLEC context

Our goal of promoting improved livelihoods by encouraging the maintenance of agrodiversity can be perplexing, but we have acquired some astute consultants who are figuring out how to put this into practice. These consultants are the small farmers themselves. Identifying and working with expert farmers has been central to our methods ever since the project transformed itself from a largely research effort to a demonstration project. Given the innovative and complex task of working with agrodiversity, finding the right farmers and the appropriate systems to promote has been a crucial focus for all groups.

Recognizing that among small farmers there are some who are distinguished by their keen interest in variation among species, varieties, plots, places and processes, PLEC scientists have sought to involve these local observers and expert investigators and to use their insights. Many of these exceptional, inquiring farmers are also particularly intrigued by the new and have a propensity for experimentation. We have put a premium on identifying those farmers who are successful because they have a record of effective innovation and change in response to shifting opportunities or problems.

The strategy is to identify diverse, dynamic and successful systems of resource management and to employ the farmers who developed and use these technologies to help test, explain and promote them. Finding these experts and systems has required a multiplicity of approaches, most of them involving extensive observation in the field. All participants and most other researchers would agree that field observation is essential to identifying successful resource management practices. However, it often can be far more tricky than expected and is never an entirely straightforward approach. What PLEC seeks, after all, is inherently difficult to see. Agrodiversity is a confusing goal as well as a moving target. But if our Clusters have occasionally missed the mark, it is rarely for lack of observation.

The importance of field observation has been emphasized in all efforts. Project personnel, even the coordinators, expect to spend a good deal of time observing agricultural patterns in the field. Even the large general meetings, where most scientific staff assemble to report on their Cluster's progress, have been held in or close to field sites. Since most visiting participants lack local linguistic expertise, the experience of these site visits has been largely a visual one.

The use of remote sensing technologies has also been encouraged, as it allows groups to take the preferred 'landscape view' (Zarin et al., 1999). Maps have been considered indispensable to much of the work (Brookfield et al., 1999). The increasing ease of capturing, storing and transmitting images has led all of us to employ and emphasize the visual in all our work. The majority of our Clusters have taken advantage of new media and produced videos of their work. Visual representation is gaining importance in all our work for this project and beyond.

Visual impressions can also be misleading. The recent work of Fairhead and Leach (1996) has forcefully argued that point, using the misinterpretation of historical changes in an African landscape as their example. While we have tried to avoid the types of assumptions about smallholder resource use that predispose to misjudging directions of change, we are certainly not immune. The emphases and methodologies direct us specifically toward the visually confusing. We have emphasized working with resource management systems that use high levels of biological diversity to achieve production goals. We have looked for systems that are quick to change and adapt. We have chosen to operate in regions where variation among villages, households and farmers is high. We have frequently opted for promotion of hybrid systems. And in many areas we have worked with minority or marginalized peoples.

Agrodiversity can be ambiguous. Biologically diverse systems are difficult to understand, especially for those trained to look for order, simplicity and uniformity in agricultural undertakings. Many of the management systems that PLEC promotes were initially difficult even to identify as managed systems. Dynamism can also be easy to misinterpret. Among environmentally concerned development personnel who now emphasize the desirability of sustainable production, dynamism can be easy to confuse with degradation. Production systems that combine traditional and modern solutions to production problems are, of course, ubiquitous but can also be misunderstood and are rarely appreciated either by those interested in development or in conservation. In addition, working among minority peoples whose languages or dialects, and particularly the unfamiliar terms they use for their management actions and systems, can also pose serious problems. An example that captures several of these problems comes from research I carried out with several colleagues in villages on the Indonesian part of the island of Borneo (Padoch, 1993; Padoch and Peters, 1993).

Perceiving agrodiversity

From the air, the territory of Tae, a village comprising five hamlets, their fields and forests, in the Indonesian province of Kalimantan Barat stands out because of its extensive forest cover. This part of Borneo has long been settled and used for production of export crops. Much of it is now deforested and planted to uniform stands of African oil palm. But Tae is different. The even-spaced, uniform rows of plantations in surrounding areas are replaced by a structural heterogeneity typical of forests. Even viewed from the ground, the dense and diverse woodlands seem to belie the known history and high population density of the region. Containing a great diversity of large trees, the forest-like woodlands of Tae give every indication to the passing scientist of being mature natural forests. Yet Tae, which covers a total area of almost 16 km^2, in 1980 supported about 88 people/km^2.

Mature forests in Tae are actually restricted to ritually conserved stands on a few hilltops; what is in abundance is instead a variety of managed woodlands that the local farmers, known as Tara'n Dayaks, make and manipulate. The various types of woodland can be distinguished by their different origins, management techniques, and production priorities, as well as by the rights of access and inheritance by which their use is governed (Peluso and Padoch, 1996). Perhaps the most interesting of the several types are those known as *mawa'n* in Tara'n, or *tembawang* in the Indonesian language.

Some Tara'n-managed woodlands begin as natural forests, some as swidden fallows. *Mawa'n*, however, originate as diverse fruit gardens. Upon building a house or making a work hut in a new rice field or rubber garden, a Tara'n villager almost invariably plants fruit trees, shrubs and herbs around the new structure. Valuable plants also sprout spontaneously from household refuse or from seeds spat out while eating fruit in the doorway. Seedlings of desirable species, whether deliberately planted or not, are often protected.

In their earliest stages, future *mawa'n* are really house gardens, some tended intensively and, like house gardens around the world, used for a great variety of purposes (Niñez, 1984; Fernandes and Nair, 1986). These areas become *mawa'n* when the house is moved or the field hut abandoned. Tara'n Dayaks tend to change house sites every 20–25 years and swidden huts are normally abandoned after two or three years. Productive fruit trees surrounding the houses are rarely cut. Many *mawa'n* continue to be managed actively and creatively by their creators, who are often remembered for many generations, and by their descendants who inherit rights to the harvest of products. A long history of occupation of the Tae area by Tara'n farmers has formed a landscape that is dotted with *mawa'n*. Today some individual mature *mawa'n* are 10 ha or more in size; in total they cover hundreds of hectares in Tae village territory. Exceptional Tara'n woodland managers claim to plant, at least occasionally, 74 different fruit species and can identify more than 100 species of fruit in the managed forests. It is both the biological diversity and the structural heterogeneity of *mawa'n* that make them 'invisible' to the visitor; they appear to be unmanaged forests or fallows. While they are not invisible to local agricultural advisers, they are not appreciated by the scientific experts who tend to dismiss or even disparage them. Visually these diverse, structurally complex, woodlands containing plants of very different origins, do not fall within accepted models of management. They are not, for example, accepted as falling into the Indonesian category of *kebun* that includes orchards and tree-crop plantations made by culturally dominant populations in the country. They are not 'legible' to those who are trained in modern agronomy or forestry (Scott, 1998).

The *mawa'n* makers of Borneo are not the only resource managers whose highly managed woodlands can be and have been easily ignored, misinterpreted or dismissed as 'bush' because of their diversity and complexity. Like the misunderstood groves of Guinea's Kissi and Kuranko people discussed by Fairhead and Leach (1996), or the diverse plots found in the Amazonia Cluster sites in Peru that are managed for timber and other products (Pinedo-Vasquez et al., 2001), forests managed by smallholders throughout the tropics, and swidden–fallow agroforestry plots around the world, have been little understood and less appreciated. Even house gardens, because of their diversity and complexity, have frequently been dismissed as messy and haphazard. These varied and intricate forms of management are not recognized because they appear disorderly and 'unmodern'. They are, like the *mawa'n*, 'illegible' to most 'modern' government officials and scientists.

Understanding dynamism

The kind of expertise that PLEC participants seek may also be difficult to spot in rural landscapes because many smallholder systems are temporally complex – they may be cyclic in nature, they are frequently dynamic, and how they develop is often contingent on a large variety of variables. Temporal complexity is frequently misconstrued. At best it is dismissed as a lack of forethought or planning, at worst it is condemned as ignorance, environmental degradation, and devastation.

Shifting cultivation or swiddening offers perhaps the most commonly observed instance of such a frequently misunderstood dynamic resource use. Until recently, phases of the cyclic system characterized by less intensive management of crops were supposed to signal abandonment of fields and agricultural activities. Now many of these dynamic systems are viewed as cyclic agroforestry systems wherein management changes in nature with each phase but never stops completely; abandonment rarely, if ever, happens.

Despite this new appreciation of swiddening, much about the nature of smallholder systems continues to confound many researchers and extension agents and makes the identification of expert farmers and expertise a demanding process. Tae again offers an illustration. Among its diverse *mawa'n* and other managed woodlands lie the various fields where Tara'n farmers produce rice, their most important food crop, as well as a number of other plants. Two main types of fields predominate in the Tae landscape: swiddens with several varieties of upland rice as the principal crop, and pond-fields (*sawah* in Indonesian) with wet-rice.[1] Among these two types, however, are plots that seem to be neither real swiddens nor adequate *sawahs*. They are weed-ridden upland fields that yield little and require extensive hoeing for production; or they are poorly watered, uneven wet-rice fields where weeds appear to outnumber rice plants. These fields seem to be evidence of Tara'n famers' double agricultural failures: a degraded shifting cultivation system and/or incompetent attempts at pond-field creation.

Only much closer, more detailed, and longer-term observation reveals that many of these apparent failures are actually fields in transition. Many of the 'degraded' sites are swiddens on their way to becoming continuously-cropped rice fields in response to a complex set of changes in the Tara'n farmers' physical, economic and political environments. Creating pond-fields is a long, complex, and labour-demanding process when earth-moving machines and hired labour are not used. The transitional stages of the process of change are themselves diverse, but often appear unkempt, unsustainable and confusing. Tae fields in transition follow different trajectories, reflecting the highly individual plans, talents and preferences of the farmers.

Tae offers a single small example of production systems undergoing change. Such misunderstood, dynamic systems are, however, found throughout the world where populations are growing, markets are changing, and government policies shift. What expert farmers do to adapt to change and seize any opportunity is often condemned as evidence of incompetence, unsustainability, and of need for intervention by scientific experts.

Temporal complexity takes the form not only of incremental change and of cyclic systems but also of concurrent management (Padoch and Pinedo-Vasquez, in press). The difficulty of detecting that smallholders who live near forests are actively managing stands of valuable trees, for instance, may be due to their often being concealed by concurrent agricultural tasks – while a plot of land may quite obviously be managed as an agricultural field by a smallholder, it may also be undergoing specific forest management operations (by the same farmer) aimed at enhancing the value of fruit trees, timbers or other tree crops in the plot. The trees that are being subjected to any number of silvicultural and other operations may not yield for another two decades, while neighbouring plants are managed for the food they will yield in a month or two.

In the estuarine floodplain of the Amazon River in Brazil, management of timber trees is routinely initiated in agricultural fields, concurrently with the management of many other resources. Simultaneity of management and multi-purpose management are among the most important characteristics that distinguish much smallholder management from the simpler patterns characteristic of industrial agriculture and forestry. Not only are smallholder fields biologically diverse at any one time, but they are temporally diverse as well. The production schedules of the various components are far more complex and require far more specialized management than do industrial-sized agricultural or forestry enterprises. This form of complexity is another one of the factors that has made field observation demanding and the spotting of potential demonstration sites, management systems and expert farmers a challenge.

Appreciating hybridity

The Amazonian timber system provides a vivid example of a final legibility or visibility problem that plagues successful smallholder production systems. They often appear to be made of disparate parts – hybrids that combine traditional knowledge and practice with modern expertise. Smallholder technologies regularly combine pieces of the local and the borrowed into unexpected wholes (Gupta, 1998). These production technologies are therefore often ignored or disparaged by the defenders of the autochthonous and indigenous, as well as by the champions of the modern.

Combination of the new and old, of the local and foreign is, of course, characteristic of all knowledge and all resource management practices. In our world of 'production packages' and of modernist (and conservationist) ideologies, however, such obviously mixed systems are sometimes not appreciated or are even belittled. More frequently they are simply not seen.

The smallholder timber management patterns that expert farmers are now demonstrating and promoting at sites in the Brazilian state of Amapá provide an outstanding example (see Chapter 17). Local tree management practices and economic activities have been transformed by several smallholder farmer-experts who participated in an earlier industrial timber boom in the region (Sears et al., 2000). A number of these farmers of eastern Amazonia have developed a profitable, vertically integrated local industry based solidly on their pre-existing agroforestry and forest management practices. These technologies continue to reflect profound locally developed knowledge of specific forests and management of ecological processes, individual observation and experimentation, but they have been updated or hybridized with knowledge and practices learned by farmers when they were employed by now-departed, large-scale, industrial timber firms. Several of the farmers have successfully combined these two ways of producing timber into a hybrid system that has proved to be an economic boon to local farmers while also maintaining high levels of biodiversity in local floodplain forests.

At each stage of the process of management and timber production, an innovative set of practices reflecting the farmers' diverse experiences is used. Floodplain forests continue to be managed, based on local knowledge of natural regeneration and growth processes, as they were before the timber booms of the 1970s and 1980s. This knowledge is now combined with more recently acquired expertise in producing those species and timber qualities that are desired in urban timber markets of Amazonia. Local labour organization, largely along kin relations, may still be used to organize labour for harvesting and processing timber, but the small timber mills and the knowledge of how to operate them largely reflect new industrial expertise and market priorities. This combination of technical, market, and ecological knowledge results in forests, timber markets and economic patterns that do not correspond to many of the widely held generalizations concerning either local or modern forest management and timber exploitation. Many modern foresters as well as biodiversity conservationists would have difficulty in recognizing the benefits of a system where timber is produced in a profusion of species, including agricultural crops and spontaneous vegetation, and where family-owned sawmills dot the rural countryside. But it is just this kind of confusing and 'illegible' system that PLEC scientists seek out and attempt to promote.

Conclusions

The concept of agrodiversity and the goal of promoting it while enhancing rural livelihoods have placed a heavy burden on project participants. While we have emphasized the necessity of being in the field and keeping our eyes open to the richness of farmers' practices, open eyes have often not seemed to be quite enough. The PLEC project has challenged us to suspend many of our preconceptions, transgress our categories, and embrace the confusing, the ragged and the illegible. Because diversity, dynamism, complexity and hybridity are all qualities of production systems that the project actively supports and promotes, workers in the field have had to endure recurring perplexity, misunderstanding and occasional failure.

Since PLEC rejects simple criteria and goals, such as promoting the most diverse system or the highest yielding variety, rapid or simple forms of field observation have often had to be discarded as inadequate tools. There are already many experiences that illustrate that, despite the difficulty, great experts and exceptional systems have been identified at numerous sites. This success has been achieved by combining skilled field observations with the close participation of many farmers; long-term research with frequent consultation with farmers' organizations. Throughout participants have refused to espouse simple dichotomies of what is indigenous and what is foreign, or simple judgements of what is obsolete and what is modern.

While PLEC has worked out methodologies that have frequently been updated and made more relevant to local conditions, no single, simple and directly replicable methods were ever demanded of all Clusters. This may seem a shortcoming to those who are used to depending on such formulations; but it is seen as a strength by PLEC. Refusing ever to become comfortable with simple categorization and simple goals, our approaches reflect what we have learned from the best of our farmer experts. Long-term success in truly sustainable production requires understanding and appreciation of bewildering diversity, incessant versatility and flexibility, and an ability to work through previously cherished categories and comfortable assumptions.

10 PLEC demonstration activities: a review of procedures and experiences

MIGUEL PINEDO-VASQUEZ, EDWIN A. GYASI AND KEVIN COFFEY

FARMERS IN A COMMUNITY in northern Thailand complained to a team of PLEC scientists that prices of the crops – cabbage and lychee – that had been widely promoted as a substitute for opium poppy cultivation, fluctuated greatly and rarely yielded them a good profit. One villager stood out. He disclosed that he did not suffer such economic ups and downs. When market prices for these two ubiquitous crops were low, he turned to his other crops: groundnuts, ginger, bananas and others. Or he fed the cabbage and lychees to his pigs, chickens and ducks and then sold or ate them. This farmer's strategy was different from that of many of his neighbours. It was, however, similar in principle, although not in specifics, to strategies of successful smallholders around the globe. He mixed his market crops with a diversity of other species and varieties of annuals, semi-annuals and perennials. The farmer managed a highly heterogeneous landscape that also allowed him to plant or manage a plethora of wild, semi-domesticated and domesticated crops at the edges of his landholdings. When less successful farmers from the village were asked their opinions of why this particular villager was doing well, they answered that 'he knows more than we do about making a living by farming'. This farmer exemplifies PLEC's 'expert farmer', a smallholder who successfully solves production problems by using biodiversity. Box 10.1 gives production details for a female expert farmer in Amazonia. Such expert farmers have been the keystones of the PLEC programme and the most fundamental resource in demonstration activities.[1]

Introduction

The PLEC demonstration activity is ideally a farmer-driven group with the conservation of agrodiversity and the improvement of farmer livelihoods as outputs. The optimum activity achieves both conservation and development objectives. This is clearly no easy task. A particular activity may require the participation of government agents, NGOs, scientists, researchers, technicians, extensionists and farmers. Bringing together such an eclectic group of participants often results in complex, sometimes contentious, interactions. The complexity is heightened because the various actors assume unfamiliar roles during demonstration activities. Expert farmers, rather than scientists or

Box 10.1   Expert farmer's diversity and income

PLEC expert farmers are chosen to teach their own agrodiverse and profitable management techniques. One of PLEC's greatest expert farmers has developed an innovative strategy to deal with market changes. She has created and perfected complex systems and techniques to increase crop diversity and other related biodiversity on her landholdings.

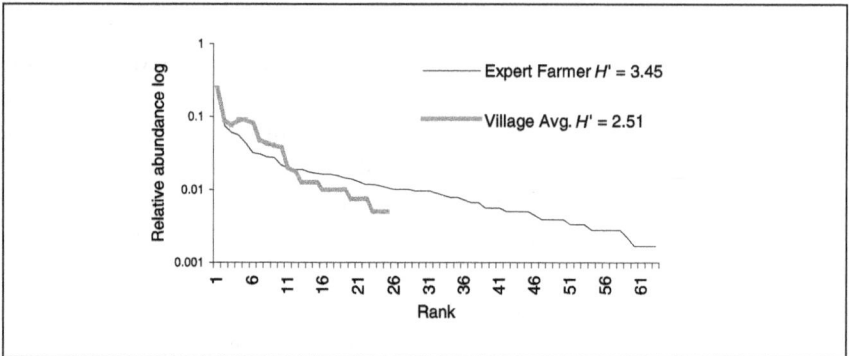

Figure 10.1   *Rank abundance in agricultural fields*

As shown in the rank abundance graph and calculation of Shannon's Index (H'), the expert farmer maintains over twice as many species in her agricultural field as her neighbours. She has profited from this diversification. From 1993 to 1999, she had an average net income over five times the legal minimum wage in Brazil. The wide range of products that she produced in her fields prepared her for fluctuating markets and allowed her to maintain a high income. Recently, the income of most residents of the village has become more dependent on timber and other forest products. Despite these trends her main source of income was derived from the sale of agricultural products, as shown in Figure 10.2. Her expertise in profiting from agrobiodiversity is a valuable source of knowledge to be shared. She has become one of the key experts in PLEC's approach of farmers teaching farmers.

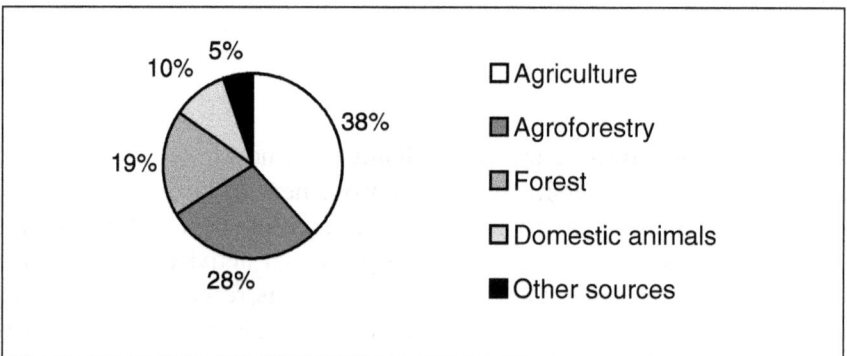

Figure 10.2   *Expert farmer's income distribution*

extension agents, supply and transfer the technical knowledge. PLEC members facilitate, monitor, observe and record this process. The other participants, ranging from policy makers to farmers, learn from the expert farmer in the hope that the new techniques or experiences will benefit their own work: sometimes this develops into a two-way process. The knowledge gained from farmer-instructors is used by farmers to enhance their production, by project directors to advance their goals of implementing development with biodiversity conservation, and by PLEC scientists to further develop our methods.

The integration of conservation and development objectives within a single project is not unique to PLEC (Agrawal, 1997; Brush, 2000). In countries where PLEC Clusters are operating there already tend to be many integrated conservation and development projects. A review of recent projects with these objectives reveals, however, that although the two ideas are often linked, they are rarely truly integrated. In most projects the union of conservation and development seldom goes beyond the title of the project and refers, at best, to the fact that a project includes separate conservation and development components. This pseudo-hybrid type of project actually offers trade-offs rather than integration. In return for following the conservation agendas of projects, local people are offered some activities designed to promote economic development or poverty reduction.

The PLEC approach differs from this currently widespread model. It is exceptional because it integrates the conservation of agrobiodiversity with activities aimed at raising rural incomes and capacity building, and it does so in a single activity. The goal is a unity, not a trade-off. PLEC uses the results of its site assessments to identify the instances where conservation and development are concurrently integrated and practised by smallholders. Researchers and technicians rely on expert farmers to identify and understand production technologies that are biodiversity-rich, economically profitable and environmentally friendly.

Conservation and development initiatives have employed many approaches for developing and transferring selected practices among smallholders (Feder and O'Mara, 1981; Phillips, 1994). A common strategy has centred upon the development of 'improved' production methods at universities or experimental stations. Extension agents then teach farmers the new methods. The ineffectiveness of this strategy, however, has encouraged attempts to reverse the direction of the transfer to make it more 'bottom-up', 'farmer-based', and '*in situ*'. Some projects have enlisted community leaders to help in this transfer, effectively delegating to them the role of extension agents (Bebbington, 1994). The results have been mixed. Sometimes farmer leaders can be effective and efficient teachers. Other projects have been hindered because farmer leaders often lack experience and may be the least successful farmers in the community, their techniques may not be replicable by the less fortunate, or the technologies that they attempt to promote were designed in research stations and are poorly adapted to local conditions (Agrawal, 1997).

A variation on the design of farmers learning from their leaders has promoted community groups, organized by gender or specialization (e.g. banana producers), to teach new technologies to community members (Uphoff, 1994). While this strategy may have helped community groups gain power, it often resulted in few real advances in the use of improved and appropriate technologies by small farmers (Phillips, 1994). There are different opinions on why the group-centred approach failed. The suitability of the techniques promoted may be questionable, as few research station designed models universally help farmers in marginal areas to deal with the complex limitations they encounter. Inappropriate techniques hinder the success of projects regardless of who transfers the knowledge. Failure to adequately appreciate differences in knowledge between members of groups has also been suggested as a reason for lack of success (Scoones and Thompson, 1994).

PLEC builds on knowledge gained from these other approaches, but is unique in its attempt to understand and use the variation among households, and to facilitate the dissemination of successful variation in agrodiversity. The teachers are not extension agents, community leaders, or cooperative farmers, but instead are expert farmers who have developed the technologies and other knowledge to respond successfully to changes in their natural and social landscapes. Acknowledging expert farmers' technologies constitutes the core of the PLEC demonstration approach. However, technologies or products developed outside the communities and adapted by smallholders to local conditions are not excluded. The approach does not try to categorize conservation practices, production systems and techniques into indigenous and non-indigenous categories. Smallholder technologies that integrate conservation and development are promoted regardless of their origin. Expert farmers are identified from among the most successful agriculturalists, agroforesters, forest managers and conservationists. Through the identification and support of these expert farmers, practices are disseminated that resolve problems resulting from changes in the natural environment, the market, or government policies.

These demonstration activities show that when expert farmers are acknowledged, their technologies and experiences are valued and quickly assimilated by the other farmers. For example, in West Africa expert farmers identified and supported by PLEC have been recognized as instructors in formal gatherings of farmers' associations, scientists, extension agents and policy makers. In Ghana, expert farmers have taken the lead in the discussions regarding new policies, community needs, and agricultural research. Many PLEC Clusters, including those in Amazonia, China and Ghana, show that even gender-based prejudices can be overcome. Several of their most active and respected experts are female farmers.

This chapter is based upon field visits to PLEC's demonstration sites. We provide a framework to elucidate the steps used by PLEC members to plan and carry out demonstration activities. We outline the kind of information required for identifying and selecting expert farmers and technologies that are promoted through demonstration activities. We review some of the results of demonstration activities and highlight some problems and difficulties encountered, as well as some of the encouraging results.

We developed a flowchart (Figure 10.3) to more readily explain the multiple steps and actors that lead to a successful set of demonstration activities. The chart shows how demonstration activities are based on knowledge gained through research into the farmers' natural and social environments. We hope that the flowchart will stimulate further modification

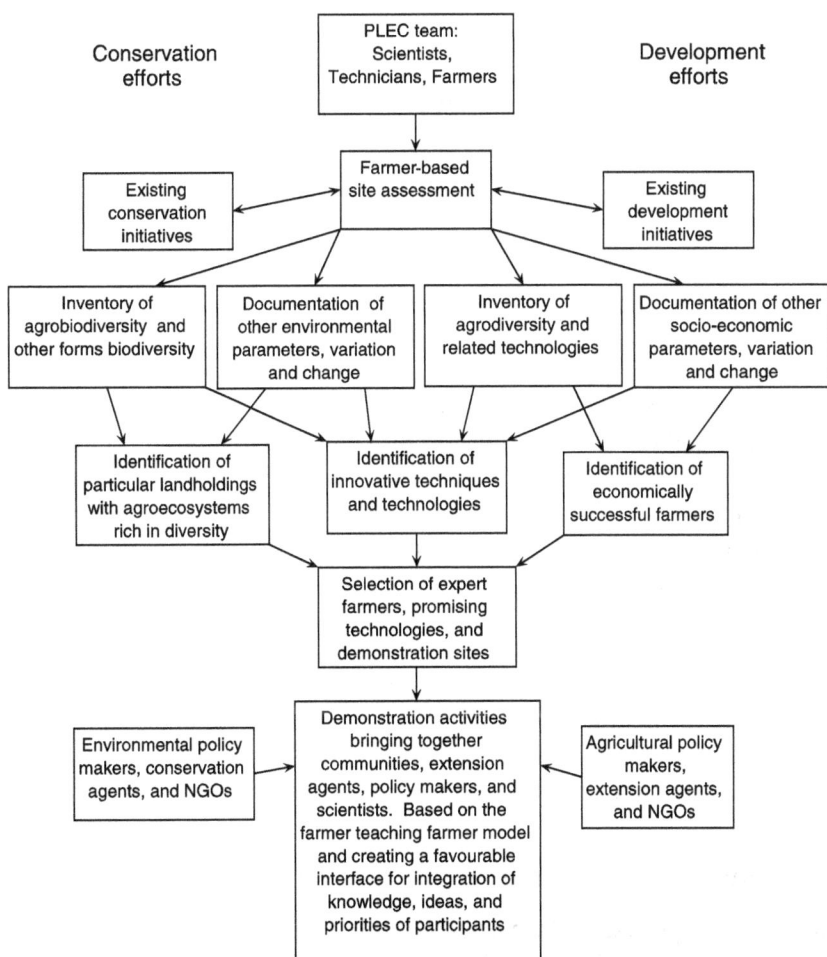

Figure 10.3 *The PLEC demonstration approach integrates conservation and development*

and adaptation of demonstration concepts to fit the many different realities encountered in PLEC sites and beyond. It should not be viewed as a strait-jacket nor followed as strictly as a cooking recipe, and is based upon a diversity of activities and approaches that can be expanded upon (Pinedo-Vasquez, 1996; Padoch and Pinedo-Vasquez, 1998; Abdulai et al., 1999; Guo et al., 2000b, Chapter 7; Kaihura et al., 2000, Chapters 12 and 14). There are many ways to achieve the goals of PLEC demonstration activities. This flow-chart is offered to provide guidance on some of the essential components for planning, executing and monitoring demonstration activities.

Assembling an assessment team

The simultaneity of both conservation and development is central to the PLEC demonstration concept. To achieve this harmonization, PLEC Clusters began their work with rigorous, multidisciplinary assessments of the variation among households of the communities at their demonstration sites. Each Cluster brought together an assessment team. The make-up of the teams has varied considerably. They vary in the number of individuals involved, the time they can spend in the field, their backgrounds and the specialized knowledge they bring to the task. All include experienced researchers and exceptional farmers. In some cases, the teams also include respected extension agents, local authorities, and religious leaders. The inclusion of external individuals must be done without compromising the basic goals of the assessment.

An important issue, which should be taken into account during the assess-ment, is the different types and degrees of outside intervention in the form of conservation and development efforts that have been, or are currently taking place in the community. To include persons closely identified with conservation or development efforts that use an incompatible approach may severely preju-dice the assessment teams' outputs. Care taken in assembling teams is always important, and will be rewarded not only in good research results, but also later when the demonstration activities take over as the central element of the work.

Local farmers play a special role as members of the assessment team. Their knowledge of the community, local production technologies, resources and landscapes is invaluable to a perceptive and reliable inventory process. The ties that the PLEC team can forge with selected farmers while doing assess-ments can also be important for the success of later demonstration work.

Farmer-based site assessment

Farmers, especially expert farmers, are at the core of all PLEC demonstration activities (Padoch and Pinedo-Vasquez, 1998; Brookfield et al., 1999). To identify those farmers and technologies that might contribute most towards the improvement of the community's development and biodiversity conservation, and to identify the needs, trends and priorities of communities, all PLEC Clusters carry out detailed and multi-focused site assessments. The multiple objectives and methodology for carrying out the assessments have been outlined in Chapters 5, 6 and 7 of this volume. Here we add only a few pertinent observations.

Many of the outputs of the assessments supply the data and create the foundations necessary for the successful planning and implementation of demonstration activities. Agrobiodiversity inventories, for instance, provide a picture of the existing variation in the level and type of biodiversity in different households' landholdings. Identifying those that maintain large quantities of rare and unusual species and varieties is a major step in choosing a demonstration activity or farmer demonstrator who can further the complex conservation and development process.

Biodiversity data are complemented with information on the performance of households as social and economic units. The household should be the primary unit of measurement and analysis. In rural areas throughout the world, decisions on how, where, when, and what to produce are usually made at the household level rather than the community level (see Chapter 7; Chapter 14; Agrawal, 1997). These household surveys reveal differences between households in a large number of crucial economic and social variables. Knowledge of both the means and ranges in income, labour availability, ownership of capital goods, and other variables, help demonstration teams identify particularly successful, flexible and resilient farmer households. Research on variations in the types of fields managed, the crops produced and the technologies employed by the households is necessary to identify those technologies that are good candidates for inclusion in the activities to be demonstrated.

Site assessments take into account a broad range of information that can also help demonstration teams understand the processes of change and the actors who participate in these changes. A variety of ethnoscientific methods can be employed, including the reconstruction of landscape histories and interviewing knowledgeable villagers. The results of this research have not only helped PLEC scientists to understand trends in local biodiversity management, but have also helped identify particularly dynamic, resourceful and resilient components of the village. These inventories provide an understanding of which of the innovative technologies that farmers are developing might be especially important in helping their neighbours cope with looming problems, or to take advantage of likely opportunities. Such technologies are then chosen for demonstration activities. Historical information on change and households' responses to change in a broad range of environmental, social and economic factors is a crucial component of the site assessment for demonstration activity planning.

The PLEC demonstration model assumes that existing biodiversity and socio-economic conditions reflect the conservation practices and technological knowledge of households. Based on these assumptions, the model integrates methods and techniques for documenting conservation practices. The body of information generated during farm-based site assessments provides a clear picture of the problems faced by households, and of the diversity of approaches adopted to deal with environmental and socio-economic changes (Figure 10.1). The third step in the PLEC demonstration model focuses on the identification of successful households, landholdings rich in agrobiodiversity, and technologies that produce and maintain biodiversity.

Carrying out assessments for successful demonstration

In reviewing PLEC's experiences in carrying out household and field assessments that readily feed into successful demonstration planning, a few common difficulties are worth mentioning. An interdisciplinary team is vital. Assessment teams that are narrow or limited in expertise will not manage to achieve the broad-based results necessary for successful demonstrations. For instance, teams whose members were mostly botanists tended to concentrate on identifying and recording lists of species and varieties present on farmers' lands, but collected limited information on the production systems or conservation practices that gave rise to this biodiversity. Teams composed largely of soil scientists focused on important trends in erosion and land degradation but left biological diversity, especially of cultivated varieties, and management diversity, under-reported. A well-balanced interdisciplinary team with clearly defined goals will ensure that assessments provide insights from agrobiodiversity and agrodiversity, as well as landscape and household surveys.

Another common limitation in the identification, documentation and selection of farmer-developed technologies stems from a mechanical or perfunctory use of categories and concepts. Some of the terms commonly used to define how farmers organize their crops, such as monocropping, polycropping or intercropping, actually reveal little that is useful about the diversity of the farming system, or how it adapts to change. Greater insights into responses to change can be achieved by dismantling the general categories and recording the technical diversity used at several stages: clearing, hoeing, ploughing, planting, weeding, protecting, harvesting, or fallowing fields. An unthinking reliance on common terms and definitions of cropping systems can lead to misinterpretation of both existing diversity and directions of change.

Integrating outputs

The outcomes of the site assessment provide many of the components necessary to plan and organize demonstration activities (Figure 10.3). Conclusions drawn from agrobiodiversity inventories, socio-economic surveys, and agro-diversity studies supply the interdisciplinary framework upon which activities can be planned. Among the important variables that need to be determined from the assessments are:

○ the crucial economic, political and environmental changes that affect land-use practices and household incomes of smallholders
○ the problems that result from these changes and how local smallholders deal with them
○ the farmers who are most innovative and successful in dealing with these problems
○ the technological diversity and specific management technologies developed by successful farmers
○ the levels of agrobiodiversity and other forms of biological diversity resulting from the application of these technologies.

The identification of exceptional farmers, techniques, and landholdings does not guarantee that an assessment team has located either the most appropriate expert farmer, or the technique that is best demonstrated. The landholding chosen for the demonstration activity will not always prove to be the best in the long term. Site assessment analysis for prospective demonstration activities should examine the households within the community very carefully to allow scientists and technicians to discover instances where successful conservation and development effectively merge. The methodology is designed to uncover variation within a community, and to identify individuals or groups of economically successful farmers who employ innovative techniques to create or maintain high levels of biodiversity in their landholdings. From among these individuals or groups, members of demonstration teams can select the expert farmers who then can be invited to show and teach these promising technologies to their neighbours in the course of demonstration activities.

Using information on management variation for demonstration activities

Translating the results of field assessments into a successful programme of demonstration activities is not an easy task. Information collected by many Clusters has, however, identified many cases that readily show how concentrating on variation and change among villagers can yield a rich store of expert individuals and expert practices that are appropriate for dissemination. A brief example from the West Africa Cluster is instructive.

Approximately two decades ago in Ghana, bush fires during an El Niño year severely damaged cocoa and other fruit species planted in the fields of small-holders. This event took place in an economic landscape where cocoa was already a failing crop; the price of beans was low, and most farmers had already experienced a rapid transition from a boom to a bust period in the cocoa economy. In the early 1990s, PLEC scientists began assessments of selected villages. They found that a majority of farmers had switched from cocoa production to a maize and cassava rotation. Some farmers, however, had responded by planting a greater diversity of crops. Others had added animal production, including chickens and even snails, to their repertoire. This variation in management activities and the concomitant diversity in economic success formed the basis for determining what effective demonstration activities might be. Researchers sought to identify what specific techniques had been developed in response to a range of economic and environmental changes, and which of these appeared to have become both profitable and biodiversity based. Answering these questions helped to find appropriate production systems and techniques for incorporation into their demonstration activities.

Identifying and documenting variation in the ways in which smallholders respond to changes such as those outlined above provides a repertoire of technologies and practices that could be used for demonstration activities. Familiarity with, and evaluation of, the practices is then necessary in order to determine which technologies or practices are helping farmers deal successfully with change, while also conserving the diversity of species, varieties, environments, and management. An experience from China provides another example. Smallholders in China were participating in state-sanctioned reforestation programmes that promote the planting of two fast-growing species. A few innovative farmers had added native species to the mix. The addition of these species was initially considered peculiar because the rotation time for their harvesting was three times greater than that of the recommended species. Through fieldwork, PLEC members found that farmers planting the local species did not need to wait until the end of the rotation to reap benefits. The native species created habitats for insects and herbaceous vegetation that favoured the growth of mushrooms, wild vegetables and the raising of chickens. In contrast, areas reforested with only the species recommended by foresters were very low in insects and did not provide varied habitats. The incorporation of the techniques used by the Chinese farmers in reforesting their land with native species is an excellent candidate for promotion through demonstration activities. It illustrates how the focus on farmer-developed practices is not backward looking and limited to traditional practices. The expert farmers are dynamic and forward-looking.

PLEC demonstration activities

Some experiences, suggestions and cautions

Over the past two years several articles about demonstration activities, expert farmers, and demonstration sites have appeared in *PLEC News and Views* (e.g. Pinedo-Vasquez, 1996; Padoch and Pinedo-Vasquez, 1998; Brookfield and Stocking, 1999; Guo et al., 2000b). Many of these articles stressed that there is considerable room for variation in demonstration activities. Building upon a common theme, each Cluster has developed distinct and evolving interpretations of what constitutes an appropriate demonstration activity for a particular situation (Padoch and Pinedo-Vasquez, 1998; Brookfield and Stocking, 1999; Brush, 2000).

Perhaps the greatest challenge has been to resist the tendency for demonstration activities to become standard development and conservation initiatives. For instance, extension agents in several sites where PLEC teams have been working are gradually incorporating expert farmer technologies into their training programmes. There is still, however, a penchant for demonstrating 'modern' or 'improved' techniques developed by agronomists and other scientifically trained experts and to have the scientist or extensionist instruct the farmers. Concerted and constant efforts must be made to ensure that activities do not merely copy conventional extension and training models. The familiarity of many participants in demonstration activities with traditional roles, and the resulting tendency to reassume familiar roles threatens the core concepts of PLEC demonstration. Based on the project experience, the most effective way to confront and overcome these biases is for researchers and technicians to increase the time they spend in the field learning from farmers.

Identification of expert farmers

Expert farmers cannot be identified without scientists having a great deal of field experience in interacting with members of the smallholder societies. Other writers active in this area, including Fairhead (1993), have also pointed out that in smallholder communities promoting exchange of production technologies should begin by identifying who knows what. Among the several characteristics that make demonstration activities different from standard extension programmes, is that this first step is emphasized. PLEC scientists take a great deal of care in identifying expert farmers by first asking who knows what.

The process of identification of expert farmers has proved to be a long and complex one. In many cases expert farmers with successful and biodiversity-rich systems of management have had unfortunate experiences with scientists, extensionists and development projects, and they have sometimes not been eager to cooperate with scientists. Farmers who in the past may have been singled out and praised as progressive farmers, and therefore recruited for multiple projects, are now not the ones who are called upon to teach. The new mode of working may prove confusing to many villagers. Experience in Peru, Brazil and elsewhere also shows that true expert farmers are often

unwilling to share their knowledge with any or all of their farming neighbours. PLEC has been careful to consult closely with farmers and allow them
to choose which technologies and which part of their technologies they want
to impart to all or some of their neighbours. The PLEC group suggests which
technologies might be of interest, the expert farmer decides which of those
she or he would like to demonstrate.

Integration of demonstration teams and delegation of responsibilities

Although the success of demonstration activities depends largely on identifying and selecting appropriate expert farmers, the composition of the entire
team and the attitude of each member towards farmers are also important
determinants of success. Based on experiences, a demonstration team can
integrate experienced field researchers, extension agents, technicians and,
more importantly, expert farmers. Where demonstration teams have been
especially successful, we find that the members developed strong relationships with the farmers. In the few instances where team members did not see
the need for establishing relationships of mutual respect with farmers,
demonstration activities appear to have had greatly diminished impacts on
the resource use of the community.

The integration of expert farmers into teams and the delegation of particular responsibilities to each member facilitate both the implementation of
demonstration activities and the establishment of demonstration sites. The
main role of expert farmers is to explain and demonstrate their production
and management techniques. It is the job of researchers, technicians and other
members of the team to facilitate meetings and activities, make appropriate
suggestions, encourage farmer demonstrators when difficulties are encountered, and monitor how farmers are adapting or rejecting the techniques
learned in demonstration activities. This is one of the reasons why members of
a demonstration team should attend every demonstration activity.

PLEC teams assist by proposing suitable and productive members of a
team. They define roles and provide some insights into what results might be
expected from demonstration activities. We have constructed a table to serve
as a guide to the possible or desirable role of team members, expert farmers,
participant farmers, extension agents and policy makers in demonstration
activities (Table 10.1). We also suggest some outputs of the activities that might
be expected from the successful performance of all participants in demonstrations. The main goal in presenting the table is to help others who might be
interested in building their demonstration activities in a similar manner.

All the components of Table 10.1 can be expanded, reorganized or modified. The matrix presented here serves only as a guideline. Several PLEC
groups have modified these specifications. For instance, in Amazonia, PLEC-
Brazil is including students as a new group of participants in demonstration

activities. At the estuarine site of Mazagao Velho, the young sons and daughters of local smallholders who are studying at a community-run school (*Escola da Familia*) began participating actively in demonstration activities over the past two years. Other groups have also rearranged, added and modified many of the activities, roles, and meeting types.

Table 10.1 Examples of activities to be included in demonstration activities, the topics to be discussed or demonstrated, the role of each group of participants and some expected outputs

Participants	PLEC team	Expert farmer	Farmer participants	Extensionists and technicians	Policy makers
Setting up the meeting	Suggests: (a) possible gathering type for activity (b) scope of participation (c) possible invitees	Responsible for selecting: (a) type of gathering (b) scope of participation (c) list of invitees	Receive invitation upon expert farmer's approval	Receive invitation upon expert farmer's approval	Receive invitation upon expert farmer's approval
Topics to be covered	Present suggestions for topics the expert farmer will cover that reflect PLEC objectives	Chooses topics he/she feels comfortable sharing with participants	Can contribute ideas for pertinent topics	Peripheral role in topic selection process	Peripheral role in topic selection process
Role during activity	Provide logistic support, facilitate interaction, and document demonstration activities	Training farmers by sharing knowledge and experience	Learn new techniques and technologies, participate in working groups and/or share experiences	Documenting and acquiring farmer's approaches to solve problems	Documenting and acquiring farmer's approaches to solve problems
Output	Defining monitoring systems and recording responses to farmer interactions	Earns respect for expertise; monetary or resource compensation; learns from interaction with participants	Incorporating or testing newly acquired knowledge in own system	Incorporating or testing newly acquired knowledge from export farmers in extension activities	Incorporating or testing newly acquired knowledge from experts in policy discussions

Establishing demonstration sites

A critical component in planning and executing demonstration activities is the establishment of demonstration sites. Based on a broad range of field experiences, the ideal demonstration sites are those set up in the landholdings of selected expert households, but only after establishing a relationship of trust with them. Several benefits are gained by initiating a partnership with expert farmers. When this is achieved, members of the field demonstration teams consult with the farmers on the best location to demonstrate a particular production or management system or technique. Team members also ask the expert farmer how many demonstration activities can be carried out, at what intervals, how many people may visit their landholdings, and other relevant questions.

Our experience and observations indicate that the varied needs, schedules and preferences of the expert and participant farmers need to be taken most seriously into account throughout the process of planning and executing demonstration activities. In addition, we recommended that the agricultural calendar be considered when planning demonstration activities.

Conducting demonstration activities

The focus on demonstration is based on the principle that farmers are always teaching and learning from other farmers. One of the most important products of this mutual exchange of knowledge in smallholder societies is indeed the agrodiversity and agrobiodiversity that we find in their landholdings. PLEC recognizes the importance of smallholder technologies and aims to promote them at the local, regional and national level by directing demonstration activities toward different social groups living in the communities and relevant regions. Not only are the disseminated techniques different, so too are the modes of dissemination. In demonstration activities villagers usually learn from and exchange experiences with farmer-demonstrators by working together in fields managed by the expert farmers. Only rarely do they learn by sitting in classrooms. Villagers who participate in demonstration activities are always free to try, change or reject the technologies that are demonstrated by the expert farmers.

As part of demonstration activities we recommend 'working expeditions', where all participants initially visit the fields, fallows, house gardens, orchards and forests owned by the farmer-instructors. These visits help expert farmers to be recognized and respected by the other members of the community, particularly community leaders. A common strategy used by PLEC members is to act as a bridge between expert farmers and the participants in demonstration activities. After a period of local activity, the level of peoples' acceptance and recognition of farmer-experts usually increases.

There are several forms of local gatherings that have been used very successfully to demonstrate particular production systems and conservation practices. Below we discuss a few kinds of gatherings that can be a basis for selecting, planning, executing and monitoring demonstration activities.

Family reunions

The expert farmer organizes this kind of demonstration activity. Members of demonstration teams help with them and note the exchange of knowledge between the expert farmer and members of the family. The exchange of seeds, seedlings and other forms of germplasm can be expected. For example, in Amazonia a woman expert farmer excels at organizing family reunions and demonstrating her techniques to her relatives, friends and others. Family reunions work very well in this case because she is the most respected member of the household and is highly regarded for her agricultural expertise. Family members consult her on what and when to plant, receive planting materials, get advice on individual problems, and learn specific techniques. Relatives come from several communities located in the region and, upon returning to their communities, further disseminate new ideas and planting materials. Their participation in family reunions has helped greatly to promote the production systems and techniques of the expert farmer beyond her own community.

Gatherings of friends and neighbours selected by the expert farmers

Organized gatherings of friends and neighbours selected by the expert farmer are one of the most efficient ways to facilitate the exchange of knowledge between expert farmers and other farmers from the community. The role of members of demonstration teams is to urge farmers to hold such gatherings and to note what is discussed among the participants during demonstration sessions and visits to the fields, fallows and house gardens of the expert farmer. An example from Amazonia illustrates one of the ways in which demonstration activities have been organized. One expert farmer is the pastor of the community evangelical church. Every two or three months, he receives members of his church and other friends at his house. During these gatherings, the expert farmer takes the group to the demonstration sites established in his forests and fallows, where he explains management techniques. A member of a demonstration team records the participants' questions and the answers of the expert farmer.

Working groups

Working groups present the most direct method for expert farmer-to-farmer exchange of knowledge. The informality and social ambience in working groups during particular labour operations facilitates the transfer of production techniques among the participants. The expert farmers organize the groups, and participants usually perform some kind of labour for the farmer. After work the expert farmer provides food for them. During or after the meal, the expert farmer explains his or her production and management techniques. The exchange of knowledge should be documented through direct observation of the working group in the fields and while the participants are exchanging comments. A member of the team can also make a list of questions and answers during the discussion between the expert farmers and the participant farmers.

Organized training or field visits

This type of demonstration activity uses methods similar to the ones used by extension agencies during their training courses. Organization of the activity is under the responsibility of the demonstration team and is conducted by the expert farmers. Based on experience, members of demonstration teams should make and use visual aids such as photographs, sketches, and posters, to help the expert farmer explain production or management techniques. In some cases the organization of these demonstration activities can be coordinated with people working in development or conservation programmes. The participation of members from other institutions can also be helpful.

An example of the use of some of these demonstration types in PLEC's Amazonia Cluster is shown in Box 10.2. It is recommended that members of demonstration teams provide as much support as possible to the participant farmers. For example, transport to the demonstration activities is often provided in Ghana and Tanzania. In the case of Ghana, team members even provide transport for small farmers from other countries in West Africa to visit the demonstration sites established in the landholdings of the expert farmers. Team members organize materials, such as photographs of activities, displays of many varieties of a specific crop, and maps of micro-environments. A visit to the field led by the expert farmer is also included in the activity.

Special circumstances and special considerations

One of the key achievements made by the research components of PLEC over the past years is a strengthened realization that many smallholder systems are essentially different from the 'scientific' or 'modern' systems that have been embraced by agronomists and promoted by agricultural planners throughout the world. Many of the smallholder technologies that are demonstrated are long-term, multi-stage, management systems where fields tend not to go through distinct stages of cropping and fallow, but where management changes year by year. The exact forms of management tend to be spatially and temporally variable and highly contingent.

Examples of such systems include the very diverse swidden agroforestry production types central to demonstrations at the Macapá sites in Amazonia, as well as many of the agroforestry systems in the China demonstration sites.

Box 10.2 Organizing demonstration activities

By working closely with the community and observing gatherings where farmers exchange ideas, the PLEC Amazonia Cluster identified several specific approaches to demonstration activities. Initially, the team planned to conduct demonstration activities as part of *encontros*, which are community or inter-community meetings where the conservation of fish and other resources is discussed. Some of the expert farmers were more comfortable sharing their ideas in smaller, less formal groups. Our experts suggested that demonstration activities be conducted using two other forms of social gatherings. The first, called *miutirao*, are shared labour groups organized by members of households to help each other with activities like making fields, planting and other production or management activities. The second type are *visitas*, which are typically gatherings of families or close friends. In all three events expert farmers are the leading figures and are the ones who invite participants to visit demonstration sites.

Table 10.3 Numbers of activities and participants at different types of demonstration gathering

Demonstration activities	Number of activities	Total participants	Average number of participants
Encontros	10	424	42
Miutiraos	54	1206	22
Visitas	34	576	17

Demonstrating such production systems is a complex undertaking since, at different stages of these multi-phase systems, the variation in management intensity, management techniques, and the resulting production, is extreme. Giving a demonstration of only one stage in the production of, for instance, fast-growing timbers in the Amazon floodplain, may not adequately characterize or describe the whole system. Taking all participants to view examples of all stages of the process may be impractical. This problem, however, tends to arise only when technicians, scientists or policy makers are included in the demonstration groups, since local farmers already largely understand the multi-stage processes. Expert farmers may, however, need to be reminded that not all members of their audiences are equally conversant with some of the long-term complexities of their systems.

Another important issue to bear in mind is the importance of minor, peripheral or edge production types, and the need to include these in demonstration activities. Although spatially insignificant, they can be locally the richest in biodiversity, including agriculturally and nutritionally important plants and animals. Edge systems may include the edges of swidden fields or bunds between irrigated rice fields, as in northern Thailand, and the agave-dominated 'fences' between upland fields in Mexico or China. Other variants have been noted at sites in Tanzania and Kenya. The economic and ecological importance of these systems makes them prime candidates for demonstration activities. In most cases, edge-cropping systems have been overlooked because they are usually small in area and have rarely been significant sources of income. Featuring them in demonstration activities may be difficult; their composition is usually highly variable, and the management apparently haphazard. They should, however, get the recognition they deserve and not be overlooked.

The economic significance of edge systems may increase greatly at times of environmental stress, such as flooding, drought or pest attack. After catastrophic floods along the upper Amazon River in Peru destroyed a large proportion of smallholder production, the PLEC team surveyed the fields and crops that had survived and began speaking with knowledgeable farmers. Some of the management systems, crops and varieties became the subject of demonstration activities which were instituted as quickly as possible. Team members continue to monitor changes and remain flexible in their determination of what is most profitably included in demonstration activities.

Monitoring the results of demonstration activities

To further develop the demonstration model, careful monitoring of demonstration activities and the responses of participants is necessary, along with a follow-up monitoring after training sessions and demonstration site visits. PLEC members have developed systems to record results for each of the four types of gatherings described. The kind of quantitative information to be reported when demonstration activities are conducted include:

- the number of training sessions and visits to demonstration sites per demonstration activity
- the total number of participants per demonstration activity
- the average number of people participating in all demonstration activities.

The monitoring of demonstration activities should include documentation of interactions between the expert farmer and the participants.

Monitoring what happens after demonstration activities is most critical for understanding how farmers and other participants are adapting or rejecting the technologies demonstrated by the expert farmers. The number of people participating in training sessions and visits to demonstration sites cannot be taken as a measure of the success of the demonstration activity. Superficial notions of 'participation' do not reveal the socio-political complexity of settings where expert farmers interact with other farmers, technicians and other rural agents. Visits to the participants' landholdings should be included as part of the monitoring process.

Information on the number of farmers who are adapting, rejecting or assimilating the technologies, immediately after and much later, is required. A great deal of time must be spent in the field during the post-demonstration period. There are several examples that show how farmers modify the production systems and techniques that they learn from the expert farmers. Results from Amazonia show that the majority of farmers do not copy the expert's techniques but experiment and modify them instead (Box 10.3).

Monitoring teams should be composed of the most experienced researchers who have extensive field experience, as well as local residents who understand the goals of demonstration activities. The composition of the team may vary in number and specialization. PLEC monitoring teams are composed of researchers, field assistants, union and religious farmer leaders and extensionists. In some cases, as in Peru, the team selects rural teachers to monitor the results of demonstration activities. Expert farmers should not be part of the monitoring team. Experience shows that some expert farmers do not appreciate the variations made by the participant farmers on the technologies shown during demonstration activities.

Although expert farmers cannot be members of monitoring teams, the participation of other farmers is critical. Experienced members and motivated farmers willing to participate are very valuable resources for a team and can document the ways in which farmers assimilate, transform or reject demonstrated technologies. In order to record the comparison between farmers accurately, the methods as demonstrated by the expert farmers must be carefully documented and thoroughly understood by the monitoring team members.

Some PLEC Clusters have recruited young farmers as field assistants for the monitoring team. The Amazonian Cluster has selected and trained two young field assistants to follow farmers who participated in demonstration activities. The assistants document the adaptation, assimilation or rejection of technologies learned from the expert farmer. These two field assistants have become invaluable members of the monitoring team. Because field assistants need to spend most of their time in the field, they are usually selected from the local villages or region. Some Clusters rely on students as field assistants for monitoring. The sons or daughters of farmers often make excellent field assistants.

A training period for all field assistants is recommended, including sessions with expert farmers and other members of the demonstration team. Field assistants are required to spend a great deal of time with the expert farmers and must be familiar with all the production technologies demonstrated. They need to be trained to perform in-field observations during the visit to the landholdings of the participants. It should be made clear that the role of field assistants is not to supervise the participant farmer, but rather to observe and record.

Box 10.3 Post-demonstration monitoring

Many farmers who participated in PLEC Amazonia's demonstration activities in 1999 began testing the techniques that they learned from the expert farmers and observed in demonstration sites. We found that smallholders were not copying the agrodiversity and other production and management techniques from the expert farmers. Instead they combined these ideas with their own and created new and original techniques. Farmers tended to incorporate learned production and management technologies only after a long process of experimentation. We found that the trial and error approach employed by farmers to test technologies and crops was increasing the diversity of technologies used through the modification of demonstrated techniques and systems. Table 10.2 summarizes the results of post-activity monitoring in Brazil. Researchers monitored the incorporation of four techniques presented during demonstration activities into the participants' cropping systems.

Table 10.2 Results of monitoring demonstration activities

Demonstrated technique	Objective	Recommended technique	Main adaptations by other farmers
1) Agroforestry banana *encapoeirada* system	Managing Moko disease in bananas	Sororoca – pariri – banana	1) açaí – banana 2) fruteiras – banana 3) madeira – banana 4) combinations of the above with banana
2) Building up soils above tide level	Production of cassava and other crops less tolerant of tidal flooding	Keep sediments and organic matter from eroding during high tides *(lançantes)* using fences	1) use of logs rather than fences 2) placing palm leaves around the highest sections of the field 3) accumulation of soils around tree trunks 4) accumulation of wood residues from saw mills
3) Enriching fallows	Production of fruits and timber	Thinning and removal of vines	1) thinning – planting 2) removal of vines – broadcasting seeds 3) thinning – broadcasting seeds 4) combining all the above
4) Managing forests	Production of fruits, timber and medicines	Removal of vines and formation of gaps *(clareras)*	1) gaps – broadcasting seed 2) removal of vines – transplanting seedlings along trails 3) gap formation – managing of seed dispersal during high tide 4) combinations of the above techniques

Community and religious leaders may also be helpful members of the team. In Peru the pastor of an evangelical church is very active in reporting how farmers incorporate production technologies and conservation practices. Community and union leaders of Ipixuna Miranda in Brazil are not only documenting the responses of local farmers to demonstration activities, but they are also reporting how farmers in other regions incorporate the technologies. Periodic meetings with these religious and union leaders are necessary to evaluate performances and improve monitoring and documenting techniques.

Conclusions

Demonstration activities are continuously evolving. The diversity of approaches and strategies being employed for conducting demonstration activities, while following the 'farmers teaching farmers' approach, are indicative of the developing process. In the short period that PLEC participants have been engaged in demonstration activities, the experience has produced valuable information supporting the tenet that poor and marginalized small farmers are holders of great knowledge, as well as developers of efficient, effective and ingenious ways of managing the world's biodiversity.

We maintain that in the process of technological exchange among small farmers, demonstration activities with expert farmers as the teachers can be an effective medium. Our experiences also show that people working in conservation and development programmes will find that, by letting farmers be their teachers, field experiences will become more stimulating and challenging. Experts can help governmental and non-governmental institutions greatly in their difficult task of finding development and conservation initiatives that reduce poverty and alleviate critical environmental problems.

11 Indigenous knowledge in space and time

KOJO SEBASTIAN AMANOR

Introduction

THE KNOWLEDGE and perceptions that rural communities hold about the environment will determine their relationship to the environment, and their capacities and willingness to engage in activities that will ameliorate environmental degradation or enable them to adapt to changing conditions. Failure to understand popular perceptions of the environment and experiences of degradation, may lead to failure in the design of 'green technologies', if these do not take local farming styles, and the objectives and strategies of farmers, into consideration. Farmers have considerable knowledge of local conditions and micro-environments. This knowledge often eludes scientists, who tend to aggregate data or to abstract models from them. Farmers' knowledge often constitutes a science of the concrete or an art of the locality. It is a dynamic system of interaction between people and the environment in which responses are continually made to changing conditions, and knowledge is continuously updated as adaptations are made and remade.

Indigenous knowledge is often supposed to be traditional – old knowledge rooted in antiquated customs and ways of doing things, which ignorance and apathy prevent rural people from overcoming. However, there is nothing traditional about the farming systems of West Africa, which have shown a dynamic ability to adapt crops from the New World and to create new varieties which have become an integral part of popular culture. This includes food crops such as cassava and maize, as well as more recent export crops such as cocoa. Farmers have themselves found ways of adapting these crops to fit into the local agroecosystems, and of integrating them into indigenous cropping patterns.

One example comes from the late Sam Asare Mate Kole, a senior agricultural officer who served at Cadbury Hall, and was instrumental in inspiring my own interest in agriculture. During the early colonial period, farmers planted cocoa 8 feet (2.44 m) apart. The Colonial Agricultural Department decried this as too close, following recommendations for a 12 feet (3.66 m) spacing worked out in Trinidad. It was, however, found that 12 feet was too wide for Ghana, and that 8 feet was the ideal distance at which cocoa would form a canopy shading the soil. The much maligned farmer, castigated as practising forms of agriculture which constituted robbery of the soil, turned out to be right. This knowledge arose not from a long familiarity with cocoa, but from the intuitive ability to apply ecological principles grasped through observation, interaction and experiments with nature.

Policy makers and government services often dismiss the ecological knowledge of farmers, and advise them to give up their 'backward ways' and follow modern methods of farming. However, modern agriculture has hardly established a good record in Ghana. Large state farms such as the Ghana Oil Palm Development Company at Kwae, and the Branam, Ejura and Wenchi state farms, stand out from the surrounding mosaic of small farms by the extent of land degradation on their enterprises.

In a recent book on the history of world technology, Pacey argues that African technology has made great contributions to modern science. He writes:

> The greatest technical expertise of many African cultures has been the survival technology they have developed to cope with the especially demanding environments. Recognition of the possibility of a dialogue between this fund of expertise and western agricultural science has come very late, after tremendous ecological damage has been done by inappropriate techniques, unbalanced development and population growth, but still offers more hope for the future than any pretence about the transfer of unmodified western technologies (Pacey, 1990: 205).

Following Paul Richards (1985), he sees that a growing appreciation of the dynamics of traditional African agriculture is beginning to shift the whole paradigm of tropical agriculture away from monocropping and 'soil and water engineering' approaches, toward a greater understanding of ecological principles, integration of trees with crops, and farming systems in tune with the ecosystem. How ironic it is that policy makers and experts in agriculture continue to decry those very principles now found to be of great value on the frontiers of science, and instead promote old and worn theories and technologies.

Indigenous environmental knowledge

Two types of environmental knowledge can be found in rural agricultural communities. The first consists of ritual knowledge, in which custom sanctions and regulates activities in such a way as to preserve the environment. This may include regulations which prohibit fishing during a particular season, which prevent the felling of certain sacred trees and plants, or which involve the protection of sacred places from cultivation or other human productive activities.[1] In the early twentieth century, the oil palm tree was considered sacred among the Krobo, and before a tree could be felled for making palm wine an elaborate series of rituals had to be performed.

While such rituals may protect the environment, no environmental consciousness is articulated with them. Thus, sacred groves are not protected specifically for ecological preservation, but often because they are ancient burial grounds or historical sites. It is for researchers to presume or project an articulated environmental consciousness on to these activities. Moreover, in the modern period many of these customs have become outmoded and no longer influence daily practice. *Akpetishie* (spirit made from palm wine) distilling is now a major enterprise in Krobo, and no one performs any rituals before felling oil palm trees for tapping palm wine.

Approach through 'custom' will therefore not get us very far. Detailed understanding of indigenous environmental knowledge must proceed from the conscious environmental knowledge that is associated with the processes of production, work and transforming the environment. In studying this knowledge, researchers need to be careful not to project their own environmental biases and prejudices. Environmental knowledge should not be equated with current global (Western) concerns about environmental conservation. It should be formulated in a broad, neutral way to embrace all knowledge about nature and ecology, which people use to achieve production goals.

Indigenous knowledge in time

In an epoch when population density is very low, and people are concerned with moving into the hostile frontier of new forest, ecological knowledge will not be involved with conservation. It will be concerned with ways of manipulating natural processes, to enable people to settle within the forest and transform it into an agroecosystem (Amanor, 1994). The types of ecological knowledge that will emerge relate to methods of farming that will minimize labour requirements, and will enable agriculture to develop while preventing rapid soil erosion, destruction of the root mat, and depletion of organic matter. The development of cocoa farming involved sound ecological principles of intercropping with shade-tolerant food crops (plantains, bananas, taro and cassava), and with an upper storey of forest shade trees. However, this took place within an overall strategy that stressed minimal labour expenditure and extensive farming practices, involving rapid penetration of the forest, and heavy use of accumulated soil fertility. As population density began to build up, frontier land ceased to be available and pressures led to land degradation. Farmers began to evaluate their farming strategies. Faced with mounting problems of weed control, declining soil fertility and build up of pest populations, some began to hark back to the old days of the frontier when real forest land existed.

For some, at least, this process of revaluation may result in a conscious effort to preserve elements of the past environment which were considered a blessing. These may include attempts to preserve certain species of trees to shade out grassy weeds and halt the spread of bush fires, to conserve woodlands, and activities to ameliorate the environment and encourage forms of forest or woodland regeneration. The negative impact of land degradation on livelihoods encourages farmers to develop new strategies based on ecological principles of which they are aware. This revaluation mirrors scientific ecological thinking. In both formal science and in popular production activities, trends may be found that lead to unsustainability, and also new trends that aim for a more balanced exploitation of the environment.

Historical dimensions

Farming practices that define the ways in which ecological knowledge is applied are likely to change in relation to transformation in the conditions of production. Farming knowledge is not static. It is important to understand the changes in farmers' perceptions and the directions in which knowledge is moving. This can be achieved in different ways:

○ Old people can be interviewed on conditions in the 'old times', how things have changed, and significant events that were associated with change. This could involve changes in vegetation, weed associations, cropping systems, crop yields, weeding workloads, availability of land and hiring arrangements, gender division of labour, labour exchange and hiring of labour, and in general how people felt about agriculture in those times. Care must be taken not to engage in romanticism of the 'old days', in producing a catalogue of how everything has changed for the worse, and encouraging informants to engage in a kind of thinking which neatly fits an environmentally alarmist ideology.
○ Interviewing young people about their perceptions of the past, and what they consider to be the significant differences may not necessarily be an accurate record, but will provide much insight into how environmental problems are being conceptualized at present, and what types of environmental knowledge are being transmitted between generations.
○ Change can be assessed through comparisons between spatially different areas that have been settled at different times and which lie at different stages in the cycle of land degradation. This can reveal much about changing patterns of environmental use and conceptualization of the environment in relation to changing conditions.

Data elicited from these research activities can be interwoven with data collected from secondary sources, including travellers' accounts, various sector and political experts, colonial archives, aerial photographs, and maps, to build up a picture of environmental change. However, the information gained from informants has the added merit of imparting consciousness and perceptions of local people to more empirical data.

Spatial dimensions

Detailed collection of ecological knowledge, classificatory schemes of nature and about particular localities carry little merit unless they are incorporated into an analytical framework. The 'my grandmother says' line of research, or 'the traditional way of doing things as recounted by such-and-such an elder' are full of interesting insights, but are inevitably parochial. Any analytical framework must proceed by documenting a wide range of experiences, and classifying and categorizing differences in production and farming strategies and concepts, in relation to material conditions. The aim must be to produce a large amount of quantifiable empirical data and information on qualitative processes at play within localities, which can then be combined in one analytical framework that synthesizes into an explanatory scientific theory.

One method of doing this is to carry out research along a transect, documenting the when, how and why of changing farming practice. This is often carried out most successfully across an ecotone where conditions change rapidly among populations with similar material cultures and extensive networks of communication. Where comparison is carried out among a more extensive sample over a wider area, it is necessary to understand the history of production in the various areas and the factors that have influenced the development of different societies. It is also important to understand the role and position of each group in the regional economy, including access and linkages to markets, social relations of production, and the history of political and economic relations with neighbouring areas and urban centres.

Field research requires formal research methods providing data on different quantifiable experiences in relation to socio-economic data and environmental conditions. This would include data on availability of land, size of holdings, type of access to land, types of soil available, division of labour, availability of household and extra-household labour, person-hours spent in farming, crop combinations, yields and farming strategies. The last of these would include tillage techniques, types of vegetation preserved on the land, weeding regimes, crop calendars, mixtures and sequences. In addition to all this, more informal data are required, revealing the problems with which farmers are coping, and the solutions they are formulating. This emerges through spontaneous conversations and visits to farms.

Whose knowledge counts?

There are no short-cut methods or magical solutions to accessing the impact of indigenous environmental knowledge, although many researchers have claimed to have developed such methods. It is necessarily a time-consuming activity. While group interviews may be a quick way of gathering and assessing information, and encouraging brainstorming, it is easy for the researcher to prompt answers in a particular direction and to gloss over serious problems which can only be articulated by individuals in relation to their own personal conditions. Ranking orders in rural societies, and concerns with prestige, seniority and authority, may also lead to silence or deference among the poor, the young and the women. The more prominent members of the community may not be those who possess the most significant environmental knowledge, and their responses cannot be taken as defining its parameters. The responses of poor farmers, young farmers and women farmers, if separately obtained, may open up new doors in the search for information. Work must proceed by talking to a wide sample of the community, including those who may not seem to be knowledgeable, and to do this on an individual basis so that respondents can exhaust all that they have to say.

Indigenous environmental knowledge emerges as the common store informing the different paths of improvisation and flights of fancy of community members. Rather than sitting under the fig tree at the chief's palace with dignitaries, it is best explored by taking off along the winding paths and discovering the extremities of the village, the chop bars with their bush-meat soup, the drinking spots, the jokers, the old women with their pithy comments, and the young women carrying water. Each of these paths reveals a different story, the diversity of which will point to common themes and precepts in indigenous perceptions, as well as to different voices and nuances.

Conclusion

Indigenous environmental knowledge refers to the modern folk-knowledge of farming people, of the relationship of their agricultural systems to the natural environment, of the various uses of natural resources, of the problems that emerge from farming, and of methods of conserving the environment and adapting to change. Environmental knowledge is conscious knowledge. It is constantly being updated in relation to changing factors of production, and changes in technology and society. It is knowledge concerned with observation, reflection, experimentation and interaction with the environment. Since it is being constantly updated in response to change and new information, it needs to be investigated through time and across different spaces, charting the factors that result in changing perceptual frameworks and processes of adaptation.

Indigenous environmental knowledge needs to be related to different socio-economic conditions, different farming strategies and responses to changing environments. An analytical framework needs to be developed which goes beyond detailed descriptions of local techniques and perceptions. The challenge is to develop a framework which merges quantitative and qualitative data, producing a wealth of empirical information about specific localities, which can be placed within the theoretical context of the relationship between ecological, technical, socio-economic and political dimensions of agrarian transformation.

12 Working with expert farmers is not simple: the case of PLEC Tanzania

FIDELIS B. S. KAIHURA

DURING THREE YEARS of working in the Arumeru demonstration sites in Tanzania, many activities have been going on among farmers, scientists, village leaders and extension workers. This chapter highlights the different experiences, and the advantages and pitfalls of working with farmers. The main focus has been on identification of expert farmers, working with them, and showing their techniques to their fellow village farmers who might learn from them but had not recognized their skills.

Expert farmers were defined in this context as small-scale farmers in local communities who more successfully and sustainably managed their farmlands despite a range of difficult conditions that included population pressure, unreliable rainfall, degraded soils and limited supply of inputs. They survived by elaborating diverse and skilful management of their environments, and had a standard of living that was above average compared with other members of the village community.

Identification of expert farmers

The project started with the assumption that farmers know their environment better than anybody else does and that survival of farmers over the years has been a result of trial and error. Continued learning-by-doing has led them to manage their environments in the way they do, and through that process they have accumulated considerable knowledge. The expert farmers are a good starting point for demonstrating to other local farmers their successful practices of resource management, and examples or explanations by scientists can complement their experience. To be able to achieve open and friendly interactions, project staff started by establishing good relationships between farmers, scientists, extension staff and the village government, and by fostering a group with the common goal of working together to improve farmers' livelihoods through sustainable use of the locally available resources.

Learning what farmers do and asking questions brought farmers closer to the project staff. They were willing to spend more time with us and even shared the secrets behind their successes and failures. Sharing meals and local beer with them after long working days persuaded the men to allow their women to participate in discussions. Usually women, particularly married women, do not talk to foreigners. Some years ago women did not even attend public meetings and, if they did, could only listen and not talk.

Initially, we asked village leaders and extension staff to identify expert farmers who could serve as the entry point for discussions with farmer communities. According to them, expert farmers were those who used such things as improved seeds, industrial fertilizers, and who planted crops in rows. On visiting the suggested farmers, we realized the wrong group had been identified. During the project inauguration meeting, which most of the NGOs and extension staff working in Arumeru attended, we also requested names of expert farmers at each site who might be suitable chairpersons to assist us in mobilizing and following up planned activities. At all sites, the farmers proposed by the meeting were later rejected in their respective villages. They were seen as off-farm workers whose success in farming greatly depended on income from outside, and they were not able to be good advisers or demonstrate acceptable practices to the small-scale village farmers. We had to start afresh, choosing different farmers. Continued discussions with farmers in the villages led to the identification of the type of experts who would be able to work in the PLEC context. It was not easy to find them. It required repeated visits to their farms, having discussions with them, observing what they were doing on their farms, and hearing from other farmers about them. Sometimes it was disappointing to hear negative criticisms of a very progressive farmer.

In general the majority of the selected expert farmers had the following characteristics:

○ They were middle-aged or elderly.
○ They usually combined knowledge of the local surroundings with experiences from outside. They had often had previous employment outside their village or district where they had taken the opportunity to learn formally or informally about new things or how to deal with problems.
○ They kept track of, or recorded, events as they happened, attached some meaning or reason to each event, and sometimes assessed the likelihood of the reoccurrence of a particular event as well.
○ They could easily predict likely events in the near future based on the current situation and their accumulated knowledge, and adjusted their actions accordingly.
○ Many of them tried different things, some out of their own imagination and others by trying what they had seen somewhere else.

In 1999 there were ten expert farmers working with the project, five at the Olgilai/Ng'iresi site and five in Kiserian. Each expert farmer addressed a different land management practice. This chapter provides an illustration of the diverse natural resource-management practices of several expert farmers in Olgilai/Ng'iresi and Kiserian and our involvement with them.

Traditional woodlot conservation and management in semi-arid Kiserian

Kiserian farmer 1, one of the expert farmers in Kiserian village, is a Mwarusha by tribe, and was born in 1926. He has lived in semi-arid environments all his life. In 1949 he moved to Kiserian in search of a larger area of land for crop and livestock production and also enough to be inherited by his children. His father was an expert in forest conservation for traditional medicines. The colonialists trained him and others in the conservation of forests where medicinal trees and shrubs grew, and they also restricted cultivation of hilltops. Kiserian farmer 1 had received both formal and informal training in forest conservation as he grew up. When he moved to Kiserian, forests were being indiscriminately cleared for charcoal making, building poles, firewood and other uses. Other farmers extensively debarked, pruned or uprooted trees in order to obtain tree products for medicinal uses.

Kiserian farmer 1, like all other Waarusha had a lot of knowledge in traditional medicine. Traditional medicine is one of the major subjects in the training of the youths when they go into the forests for circumcision. Some of these medicines are used to treat them after circumcision. He noted that most medicinal plants and trees had disappeared in Kiserian.

At the time of the assessment, his farm had four Field Types. These were maize with beans, chickpeas, natural pasture and conserved woodlot. Unique to his farm was the conserved woodlot. It was a clear example to other farmers of biodiversity conservation in a semi-arid degraded environment. The woodlot had the greatest diversity of trees, shrubs and grasses in Kiserian. In a 100 m^2 area of the conserved river ravine woodlot 57 species were recorded. Most trees and shrubs had various uses but these were mainly medicinal. The bark, roots, leaves, fruits or seeds of 30 species had medicinal uses. In total the species in the woodlot had more than 300 uses. There were 22 fuelwood species, 13 were nitrogen-fixing species, 13 species were used as fodder, three were used for food and nine had no designated use. The most abundant plants were the multipurpose trees *Albizia petersiana* and *Combretum molle*. Most of the trees were natural but some had been collected from other places to add the trees providing the missing economic and social values. Some of the added tree species were those considered by the farmer to be endangered due to excessive use. He had trees planted for timber as well as firewood. He also obtained building poles and had erosion controlled in all places where trees were growing. He says he has fresh air around his homestead, which is close to the woodlot, and claims to have more rain than elsewhere in Kiserian where the land is bare.

The woodlot became an island of attractive forest where other farmers, without forest, but in need of building poles, medicines and firewood, occasionally encroached. The farmer made the following by-laws for his own household and the villagers regarding the use of his woodlot:

○ Nobody is allowed to cut a single tree from the woodlot unless permission is obtained from him personally.

○ Nobody is allowed to collect firewood from the woodlot without his permission.

○ All individuals in need of medicine from the woodlot will do so with his permission and supervision.

○ No person is allowed to bring animals into the woodlot for grazing. All animals found grazing in the woodlot will be sent to the village administration and owners will be fined.

○ All individuals interested in propagating some of the trees that are conserved in the woodlot and are no longer available elsewhere are advised to consult him for seed, cuttings or any other propagation material.

○ It is illegal to trespass, fire or harvest anything from the woodlot without permission from him.

Such by-laws made him unpopular with most of the villagers. The village leaders also did not help him enforce his by-laws in that no immediate action was taken against trespassers. At the time that we became involved, the farmer had disputes both with neighbours and the village leadership. We immediately recognized his efforts and identified him as an expert in environmental and biodiversity conservation, and praised him for what he was doing. Praising him alone was enough incentive for him to join in demonstration activities. During our discussions he asked PLEC to educate farmers of the need to conserve and plant woodlots and stop indiscriminate abuse of the already conserved areas. In supporting him, PLEC organized several field days at his farm where farmers from inside and outside Kiserian village, the village leaders and NGOs working in Arumeru were invited. Kiserian farmer 1 was then asked to educate the audience on the importance of conserving woodlots and biodiversity. In addition, we urged the village government to encourage people like this farmer to continue conserving biodiversity, and for farmers to visit his farm and learn from him.

The need to make use of local farmer experts in training other farmers on the best practices of resource management has always been emphasized during demonstration site discussions. Some of the NGOs and extension staff working in Kiserian had previously always passed by this woodlot without paying attention to what the farmer was doing. They thanked PLEC for opening their eyes. They also initiated arrangements with Kiserian farmer 1 to bring farmers from their projects to his woodlot for training. Several opportunities for him to address different farmer audiences in workshops and meetings were arranged. He has gained popularity in his village and outside, although there continued to be a few neighbours who envied his progress and recognition. Kiserian farmer 1 was rewarded with a wheelbarrow, a hoe, a machete and a spade for outstanding biodiversity conservation in semi-arid environments.

Kiserian farmer 1 has always been ready to share his experiences and knowledge with other farmers through farmer field days and village meetings. He has managed to convince some of his neighbours, especially those trespassing in his woodlot, to plant and conserve their own woodlots. He has always been willing to supply seeds to those in need. He plans to hand over the woodlot to his children and move to a new place where he will establish a new nursery for selling seedlings and start a new woodlot.

Intensive agriculture in densely populated environments: Olgilai/Ng'iresi

At the Olgilai/Ng'iresi site, land pressure is one of the dominant constraints that farmers have to live with. Most farmers have less than 0.5 ha of land for multi-purpose uses. Most of them demonstrated a high level of agricultural intensification as a survival strategy to avoid out-migration. Ng'iresi farmer 1 is an expert farmer of agricultural intensification and agrobiodiversity conservation. He previously worked as a watchman in Arusha town. He has attended several rural development seminars, participated in various farmer outreach programmes with some of the NGOs working in Arumeru and attends annual national agricultural shows where he collects different materials as inputs to his small farm. He has a habit of collecting different plant materials that he considers to have social, economic, cultural or ornamental value from the different places he visits and plants them on his farm.

During one of our visits to his farm he had the following to say: 'My parents were once on a small piece of land in this village. They failed to earn a living and couldn't feed us. We moved to Kiserian and that is where I grew up. However, I was personally not interested in staying away from my ancestors' land. I decided to come back to Ng'iresi and bought this small piece of land. For the past nine years I have tried different things and managed to support my family using this small farm. I do not plan to move to any other place. Some of the practices that have proved successful here will be introduced in my support farms a few kilometres from here. My children then might think of moving if they wish'. During a visit to his farm in March 1999, he had twelve Field Types each with a different cropping system and a complex mix of crop varieties. There was diversity in planting times and application of inputs.

The farm was a model for intensive agriculture in densely populated environments. Within the 12 Field Types, there were ten different food and cash crops, six types of trees, more than ten medicinal plants used for curing more than 30 diseases, 17 types of nursery seedlings for propagation and sale, six vegetable crops, 18 fruit trees, and seven ornamental plants. These are outlined in Table 12.1. There were more than 45 varieties of crops, vegetables and fruits, each with unique characteristics. The plant species on this farm of less

Table 12.1   Plants grown in the Field Types of one expert farmer in Olgilai/Ng'iresi, Tanzania

Field type	Area (m²)	Crops and varieties
1. Potatoes and maize	1078.3	Rongai potato (introduced high-yielding variety with low market value). Kilima maize (introduced early high-yielding variety with good milling and high value)
2. Beans and sweet potato	960.63	Maasai red bean (early high-yielding climbing local variety of *Phaseolus*). Burka sweet potato (large white tuber, disease and drought tolerant)
3. Boundary	74.20	*Grevillea robusta*, *Calliandra calothyrsus*, neem (*Azadirachta indica*), *Alocasia macrorrhiza* (2 local varieties taro), *Dioscorea* (local yam variety with large tubers and high value)
4. Matatu	2345	Coffee (1 introduced variety and 1 locally bred more disease-tolerant variety). Banana (7 varieties with different fruit qualities and disease tolerance). Potato (West Kilimanjaro, early and high market value). Maize (Matatu, bred by the farmer by crossing with introduced hybrid varieties). Cauliflower
5. Young coffee	465.4	2 varieties of banana with maize (Kilima), beans (Maasai red), *Grevillea* and *Calliandra* with coffee
6. Wet season garden	188.27	*Solanum nigrum* (green vegetable, local and introduced varieties). *Cyphomandra betacea* (tomato tree eaten as fruit), *Citrus nobilis* (tangerine), orange, cauliflower (seedling nursery), cassava (lasts up to 3 years in ground), banana, *Withania somnifera* (medicinal)
7. Dry season garden	153.55	Local passion fruit, *Annona squamosa* (custard apple), tree tomato, *S. nigrum*, guava, arabica coffee
8. Mseto	740	Banana (4 varieties, one different from FT4), taro, maize (Kilima), beans, *G. robusta*, *S. nigrum*, passion fruit, *Dioscorea* yam, local variety of *Annona squamosa* (custard apple), pomegranate and avocado (2 local fast-growing varieties, one producing after 5 years)
9. Man-made soil	71.44	*Croton macrostachyus*: fast growing for leaf litter, nectar for bees, timber, windbreak, shade, firewood, support frame for climbers, erosion control and medicinal uses. Passion fruit, taro and *Dioscorea*
10. Medicinal	466.65	9 species
11. Nursery	25.6	Avocado, *G. robusta*, *Cynodon* spp., rosemary, 2 *Aloe* species, coffee, *Pinus patula*, *Bougainvillea*, *Helichrysum*, *Asparagus racemosus*, rose, *Leucaena leucocephala*, *Cupressus lusitanica*, *Carica papaya*, *Tephrosia vogellii* (nitrogen-fixing)
12. Fallow	300	*Pinus patula*, *Cupressus lusitanica*, *Pennisetum clandestinum* (kikuya grass), banana, *Cucumis anguria* (calabash), pumpkin, *Olea welwitschii* (multi-purpose tree)

than 0.5 ha had more than 60 diverse uses. Interesting, however, was that the farmer also had three dairy cows but no grass. A few trees provided fodder, and through intensification of cropping he earned surplus money to obtain grass from other people.

It is important to note that some crop varieties were selected and planted for their aromatic, spicy or sweetening characteristics when added to food. Mshale laini, for example, is a banana variety which gives good taste with or without being mixed with meat. Maasai red is a bean variety that sweetens food when mixed and cooked with bananas or maize to make *makande*. Use of locally available resources not only provided inputs like manure and compost, but also provided products for making foods delicious without using ingredients from outside the farm.

Due to land scarcity, Ng'iresi farmer 1 had collected household refuse and crop residues and established man-made soil (Field Type 9) where he planted different types of yam, taro and passion fruit that climbed on the *Croton macrostachyus* tree. He said that he was able to feed his family with yams harvested from this man-made soil for one month. He had three of his Field Types planted on plots borrowed from neighbours who did not use their land. They had probably moved to other places without selling their land. Unlike Kiserian farmer 1, whose medicinal plants mostly cured animal diseases, Ng'iresi farmer 1 had more medicinal plants for treating human diseases.

After he had been recognized as one of the expert farmers, he was asked to join demonstration activities and train other farmers willing to learn from him at his farm. He organized two demonstrations at his home, one on the production of potatoes and another on maize production. He participated in five on-farm demonstrations conducted by other expert farmers in his village, and visited three expert farmers from outside Ng'iresi. He participated in locally organized workshops twice, and at a national workshop he presented his experiences in farming in Arumeru. All of these were organized by PLEC.

During his demonstrations, attended by farmers from his own and neighbouring villages and from Kiserian, farmers exchanged knowledge on potato production, extensively discussed his successful practices, and some of the participating farmers demonstrated even better practices than he was using himself. Two-way contributions between farmers combined with complementary inputs by project scientists formed a powerful learning tool that enhanced rapid adoption of what each farmer individually considered relevant or useful in their own particular farming conditions. As indicated previously, Ng'iresi farmer 1 had a collection of plant materials from different places. Farmers visiting his home during demonstrations collected several of his plant materials to grow on their farms. He in turn collected material when he attended other demonstration-site training. In this way farmers shared knowledge and exchanged planting material. PLEC facilitated these exchanges as a mechanism for conserving biodiversity among small-scale farming communities.

Training visits have become a routine activity which farmers demand when there is delay in arranging them. Discussions at the demonstration sites have gained momentum and are attended by non-PLEC farmers and even farmers from other villages. It is, however, difficult to measure the extent of adoption of new practices by both expert farmers and previously poor land managers, or to measure the spread of knowledge gained during demonstration-site discussions especially by non-PLEC farmers, NGOs and other institutions. Farmers and extension staff have continued the exchange process. Recent demonstrations were organized and conducted by farmers and extensionists by themselves without PLEC staff involvement.

Farmers have identified the benefits they have gained through working in demonstration activities. In recent discussions, Ng'iresi farmer 1 communicated the following benefits:

○ exchange of knowledge in crop and livestock production with farmers and experts from different parts of the world visiting his farm
○ visiting other farmers enabled him to learn a lot about agriculture and environmental management
○ PLEC Arumeru presented him with a wheelbarrow, hoe, spade and machete, which he didn't expect to get
○ his wife and other women in Ng'iresi benefited from the poultry project that was initiated in all project villages by PLEC
○ he learned and still continues to learn more from fellow farmers on the importance of planting trees for soil and water conservation as well as for different uses.

Vegetable production and organisational diversity in Ng'iresi

Ng'iresi farmer 2 demonstrated expert vegetable production practices and successfully allocated different Field Types to different parts of the farm. Variation of Field Types was dynamic and changed with changing seasons and years. Variation is usually an option available to rich farmers with more land, but not for poor farmers who have hardly any land. The organisation of crops and cropping systems by Ng'iresi farmer 2, who has an average sized farm in Ng'iresi village, was recorded for the period 1998 to 2000, and is briefly described in Table 12.2.

Ng'iresi farmer 2 had less diversity of crops and crop varieties than Ng'iresi farmer 1, who was of the rich category. The farm contained five food and cash crops, one vegetable crop, one type of fruit, five types of trees which were mostly boundary and fodder trees, and one grass species. He had five banana varieties. Ng'iresi farmer 2, on less land, had more changes in crop varieties within the perennial crops with the changing seasons than Ng'iresi farmer 1. Some crop varieties were maintained every season. During the short rains in

Table 12.2 Crop sequences during six growing seasons on one farm in Olgilai/Ng'iresi, Tanzania

	Area (m²)	Field Type long rains 1998	Field Type short rains 1998	Field Type long rains 1999	Field Type short rains 1999	Field Type long rains 2000	Field Type short rains 2000
1	50	Boundary with multi-purpose trees	Unchanged	Unchanged	Unchanged	Unchanged	Unchanged
2	790	Maize and beans	Fallow	Potato, maize, cabbage	Maize, beans, sweet potato	Fallow	Sweet potato, beans (Massai red), maize (Larusa)
3	1584	Potatoes/coffee	Coffee, bananas and maize	Coffee, banana, maize/beans and potato	Coffee, banana with beans	Potato (W.Kilimanjaro)	Maize (Kilima)
4	3675 (1800 and 1875)	Pasture (Kikuya)	Pasture; and a part ploughed to become field and agroforest	Pasture; and cabbage and maize intercropped with guava and Sesbania	Pasture; potato (Rongai), and sugarcane added to guava and Sesbania	Pasture, potato (W. Kilimanjaro), with potato (Rongai) under trees	Pasture, potatoes, perennials unchanged
5	532	Agroforest of coffee bananas and citrus, with maize (Kilima) and beans (ngichumba nado)	Unchanged	Trees unchanged, maize (Larusa), beans (Maasai red)	Trees unchanged, maize (Kilima), beans (ngichumba ngiro)	Trees unchanged	Maize beans and potato to be planted
6	84	House garden: *Solanum nigrum*, kale, *Capsicum frutescens* (chili), ngogwe (bitter tomato), spinach	*Capsicum annuum*, *S. nigrum*, ngogwe	Papaya *Amaranthus hybridus*, *S. nigrum*, spinach	Kale, *S. nigrum*, pepper, ngogwe, spinach, papaya, *Datura* flowers, *Amaranthus*	*S. nigrum*, taro, kale, spinach, papaya, hibiscus, *Amaranthus*	Beans, kale, *S. nigrum*, pepper, ngogwe, spinach, papaya, *Datura*, taro, hibiscus, *Amaranthus*
7	243	Grass fallow	Grass fallow	Cabbage	Grass fallow	Potatoes (W. Kilimanjaro)	Potatoes (rongai)

1998, the pasture plot was converted to perennial crops. During the long rains 1999, he added cabbage and potatoes to the perennial crops and started planting grass and fodder trees on the boundaries and contours. There was also a rotation in varieties of annual crops, especially maize and bean varieties, with each season. Rotations ensured the balanced removal and input of nutrients into the soil, reduction of pest and disease incidence, changes in family food combinations, response to market demand, improved productivity, and tolerance of different kinds of environmental stress. The integrated system reduced the need for the farmer to apply external inputs, but he still obtained a sustainable harvest. Besides demonstrating professional vegetable production to other farmers, Ng'iresi farmer 2 contributed significantly in demonstration-site discussions. He also visited other farmers in his own time for further individual discussions on recommended practices.

PLEC and the land-poor

With more land, a farmer has greater opportunities for diversifying crops and cropping systems than under conditions of land shortage. Ng'iresi farmer 3 is a poor farmer with a very small piece of land (<0.04 ha). She depended on her elder son's income generated through working in town, and also spent a lot of time picking firewood from the forest for sale. She had one cow, a sheep and three chickens. Many production problems limited her crop yields, including coffee berry disease and vermin.

Ng'iresi farmer 3 was not interested in the project initially as she considered attending discussions at demonstration sites to be a waste of time. However, project staff frequently visited and discussed alternative ways of better managing her environment. She was invited to present her farming situation to a large audience at one of the workshops. At the workshop, she received a great deal of input from the audience on how to manage using a very small plot. Most of the advice came from farmers she already knew but from a different venue. From then on she became one of the most enthusiastic PLEC farmers. Although her cow was still a heifer, she was looking forward to drinking and selling milk when it calved. Ng'iresi farmer 3 also wanted to grow more different crop varieties on her farm, although she was constrained by lack of space to easily mix annuals with perennials. In 2000 she planted *Setaria* grass on the boundary to increase feed for the cow.

She had only two Field Types: an edge or boundary and a mixed cropping plot. There was little change of crops and cropping systems in different seasons and years and she had few improved varieties. She concentrated on four food crops and four vegetable crops with some pasture, with beverage and medicinal plants on the edges or boundaries. There were a total of 18 varieties of crops and vegetables (all local varieties). Compared to Ng'iresi farmer 2 (average farmer) who had 26 plant varieties on 0.7 ha of land, Ng'iresi farmer

3 with 18 plant varieties on 0.04 ha of land had high biodiversity and attempted to spread the risks of crop failure. Farmers like Ng'iresi farmer 3 require more assistance, and many follow-up visits were made to ensure they did not miss demonstration-site discussions.

Problems associated with a farmer's efforts to conserve agrobiodiversity

With continued efforts to identify expert farmers in Kiserian, project scientists visited a farmer who had been somewhat neglected by other villagers. Kiserian farmer 2 had worked for colonialists in gardens and horticultural farms and had even visited different farms in Uganda, Kenya and Rwanda. He had been a cook for colonial staff. He then developed an interest in tree planting. His farm was a mango-tree-dominated agroforestry system. He collected different varieties of mangoes and other trees and planted them on his farm, travelling near and far in search of disappearing or endangered trees. There were several trees planted to improve soil fertility, and some of the rare sweet potato varieties previously grown in the area were found only on his farm.

His land was green and shady most of the time. Neighbours came there and grazed livestock, which destroyed his trees and crops. For this reason he tried hard to plant trees for neighbours. By the time the trees were firmly anchored, landowners uprooted, set fire to them and destroyed them, including some on his farm. They said that Kiserian farmer 2 was planting trees on their land in order to take over the land when the trees established. He had several disputes with neighbours, and many accusations were made by the same neighbours at the village head office. Some unknown villagers also strangled and killed his wife when she had visited her parents in a nearby village. When PLEC identified him as an expert in agroforestry, he was already a desperate man and isolated by the community. Neighbouring farmers and the village government did not even want us to go to his place as they associated him with witchcraft. At that time one of his daughters had been denied registration for primary school education. We were warned not to go near that 'bad man'.

We drove to the front of his house and he came and received us warmly. He thought we were the white people who used to visit him to buy fruits and vegetables and find him a market for his farm products. He was moved by our decision to visit him after years of being isolated by neighbours and pleased to hear that we had been attracted to his farm from a distance and wanted to learn from him about what he was doing and how he managed to succeed while others were failing. Our visit was a relief to this man who had not been recognized for a long time despite practising environmentally friendly farming. He gave us some roasted sweet potatoes to eat then showed us around his farm. He also told us about his bad relations with neighbours and the village government. Kiserian farmer 2 lamented that his neighbours could learn better things and earn money, and instead they came to steal fruits from him. With good rains, he was able to sell five trucks of mangoes in one season. However, he had very few friends who recognized his success or visited him in search of advice or planting material for their own farms.

After long discussions, we told him that he was one of the best farmers in Kiserian, and that we wanted him to teach others how to conserve their environment better. We asked him to help other farmers to be able to harvest something during the dry season as he did. We promised to consult the village government and discuss his problems with them. As we had come into the village with a blessing of the village government our discussions with them were successful, and they promised to convince neighbours not to isolate this farmer and to attend on-farm discussions that were organized at his home. The farmer became less angry with those who considered him a witch. His daughter registered at school after we asked the primary school head teacher. With all these things Kiserian farmer 2 became a very strong ally of PLEC. He made visits outside Kiserian to see what other farmers were doing. PLEC also supported him by purchasing barbed wire to fence and protect his endangered tree collections from being grazed by neighbours' livestock.

Older people have more accumulated knowledge than the younger generation. Besides being an expert agroforester in semi-arid environments, Kiserian farmer 2 (an old man) was the only source of information regarding soils and their local nomenclature and uses. He gave a list of names of local soil types and their occurrence within the local physiography (see Chapter 14). Armed with this knowledge, experienced farmers are able to select the appropriate soil types for growing particular crops. They sometimes made some modifications such as terracing or ridging to increase crop and soil productivity. Due to high population pressure and the scarcity of productive land in Kiserian, villagers were beginning to use *mbuga* soils (previously used for grazing) for crop production. PLEC also conducted demonstration-site discussions that involved primary school children as a way of encouraging the transfer of local knowledge from elderly farmers to the younger generation.

Conclusions

These examples of using expert farmers to educate fellow farmers on good practices offered the advantage of bringing about rapid changes to small-scale farmers' methods of managing resources and improving their livelihoods. However, there are potential problems that may be associated with working in villages if care is not taken. Through experience, project staff in Arumeru learned that before every step it was necessary to make a careful analysis of the likely outcomes. Some of the potential problems are outlined below.

o Village government support may be lacking if, during the initial stages, the selection of farmers does not respect village government regulations, or if due respect to the elders is not clearly demonstrated by researchers and extension workers.

o There may be competition with other projects in the area. This has sometimes been the case elsewhere where different projects compete in the same villages for the same farmers. We avoided competition by inviting other projects working with farmers in Arumeru to attend the PLEC project inauguration meeting in 1998. At this meeting the objectives of PLEC and the methodology were explained. Complementary activities that could be undertaken in a collaborative way were identified. Cooperation with the various organizations continued by inviting them to farmer field days and workshops.

o Involvement in politically sensitive matters or village organization, like the cases of the two Kiserian farmers, may cause problems. Involvement in such disputes, though made with good intentions, could have resulted in turmoil if farmers and the village leadership had interpreted our actions negatively.

o Often the best times to meet farmers were when they had serious problems. In Kiserian, for example, identification of interested and willing farmers came when the village was under quarantine due to a cholera outbreak. Since this was the only opportune time to easily meet farmers, project scientists had to obtain permission to work in the village and had to take preventive drugs. This decision also made the village government understand that we were determined to build close collaborative relationships. They continued to accompany us in the field whenever they had time.

Failure to meet farmer expectations may limit success. Initially farmers wanted to join PLEC activities because they expected financial benefit, as is the case for many other projects. However, after being told that our collaboration and work was limited to information and knowledge sharing, farmers gradually educated each other and became involved, with a clearer picture of what to expect. Now they say that a person who gives you knowledge is a thousand times better than the one giving you money because one can do so much more with the knowledge. In Kiserian, farmers had asked PLEC to assist first in demonstrating methods of moisture conservation and fertility improvement. The first year of successful crop performance, using the methods we proposed, made farmers realize that we were not taking chances but were sure of what we were doing. If the initial demonstrations had failed, then PLEC would not have been successful in Kiserian.

13 Genesis and purpose of the women farmers' group at Jachie, central Ghana

WILLIAM ODURO

Introduction

JACHIE IS ONE of the small rural towns (or large villages) that characterize the Ashanti region of the dry semi-deciduous forest ecosystem of Ghana. Although this ecosystem suffers from severe land degradation, remnant patches still remain that remind us of the past. Farmlands at Jachie are either tenant- or owner-occupied. In the recent past, farming has been destructive of the vegetative cover, leading to accelerated soil erosion and reduced agricultural yields. In order to overcome these problems, the PLEC project was launched to improve biodiversity conservation. The process of biodiversity conservation was accompanied by development of village institutions that have begun to take up the task of managing natural resources on a sustainable basis.

At the beginning of the project, it was realized that any attempt to stabilize the condition of farmlands in the Kumasi area would not succeed without the active involvement of women. They are the main users of natural resources in Jachie and many other villages, and they understand the environmental problems. This chapter describes how Jachie women farmers became involved in biodiversity conservation activities. The approaches adopted, the purposes, the challenges faced and progress made by the project in this direction are also presented.

Gender issues at Jachie

In this part of Ghana, women are both the principal farmers and also the main users of forest and common-land resources. Men from this village migrate to the nearby city of Kumasi to earn money more quickly, leaving the women as heads of households. Seventy-four per cent of farmers in Jachie were reported to be women (Oduro, 1998). The mean age of the women was 42 years, ranging from 22 to 85 years, while the mean number of years of farming experience was 19 years, ranging from 1 to 75 years. Seventy per cent of the women were married and they had an average of five dependants. Most of the women had completed middle-school (48 per cent) or non-formal education (32 per cent). The highest educational level achieved was secondary/vocational (10 per cent). Fifty-eight per cent of land was self-owned.

Absence of the men had substantially changed the balance of rights and responsibilities, and terms of resource use and management between men and women. A new spatial division of labour has been created, forming split households with rural roots and urban branches. Women farmers have increased responsibility as the caretakers and sources of knowledge of the ecosystems that provide food, fuel, water, shelter and the cultural foundation of their community. Women have incorporated knowledge, work and responsibilities that were formerly part of men's domain. Now they are leaders in the conservation of agrodiversity.

There is a range of products gathered by women from forest land (Falconer, 1992; Townson, 1995; Oduro, 1998). These products are used in their daily lives or are processed into marketable commodities. In the case of Jachie women, trees are the source of a variety of products and their use is well integrated in their daily life. The relationship between trees and the women was the starting point in planning and implementing the biodiversity conservation project at Jachie and its surrounding communities. Some forest products, such as snails, mushrooms, fruits and leaves, form an integral part of the household economy (Oduro, 1998).

Obtaining adequate fuelwood for domestic and other economic activities was a major problem. To avoid further deforestation, restore degraded areas and improve crop yield through soil fertility enhancement, measures needed to be taken to control and sustain the current level of fuelwood production.

Approaches adopted in Jachie

Initial contact with PLEC was made by the Jachie women themselves, in particular by the charismatic lady, Ms Cecelia Osei, who has from the beginning been a driving force in the local organization and its work. Project scientists made several trips to the village and held meetings with various groups, including chiefs and elders, and assembly and village development committee members. The meetings, although chaotic in nature, brought a level of a rapport between scientists and the village community and created awareness and community mobilization. We undertook extensive field trips in the project area, spending time with the village women in their houses, on their farms and at funerals and weddings. These visits helped in understanding the village dynamics and some specific issues that relate directly to women, such as child marriages, single parenting, and multiple marriages and breakdowns. It was during the initial interaction with the village women that it was learned that the cultivation of food at subsistence level fell mainly on the women. Apart from their daily domestic and farming activities, women keep small numbers of livestock, such as local fowls, ducks, goats, sheep, and occasionally collect plant medicines for trade to supplement their incomes.

There was awareness among the women of the need to transform and improve their socio-economic situation. No previous attempts had been made in Jachie to get the women together; and 'organizing' was an unknown word until the project team decided to work with them. We sought people who were selfless and ready to share knowledge and biological materials. We had to live with them to learn what they know. We discussed all plans with them before implementation, and began from what they know, using what they have, and approached the farmer with respect for her intelligence, ambition and tradition. Once the scientists had learned from the farmers, they then built upon the farmers' knowledge by introducing scientific knowledge. The scientists always responded to the women's needs and this attracted the women to the project. However, we also got them to chip in as part of the bargain in order not to create a dependency situation.

At a subsequent round of small meetings, the concept of identifying leadership was discussed. Participants at these meetings were from different socio-economic backgrounds, with the majority from economically weaker sections of the village and female-headed households. There was a spirit of cohesiveness in these groups. A few who had good communication skills were willing to accept the responsibility of organizing women's groups and soliciting participation. An association of women farmers with popularly elected executives was inaugurated in a community forum. On completion of this process the foundation of the work was laid.

The participatory approach had a number of key features:

○ The initiators of the project at the village level were local women who were in close contact with the target group and with whom development of a close and trusting relationship was possible. Quite a few of them already had certain standing within the local community, and this fostered quicker bonding of the new relationship.
○ No bureaucratic procedures were involved and no membership limitations regarding age or educational requirements were set, so as to avoid excluding any local village women.
○ The leaders formed a link between the scientists and village women.
○ Financial costs were much less than having the same tasks performed by non-villagers.

The leaders were given support in learning organization and communication skills. The training played a crucial role in improving job performance by developing knowledge, skills and attitudes. The women leaders then conducted village-level meetings to increase awareness about project activities.

The purpose of the women farmers' group

As the project gained experience in working with the women, many new and innovative measures were introduced that were specifically designed to increase agricultural diversity and the participation of more village women. The initial approach was to establish a model outreach experimental farm on the land of one of the leading members, that would meet most of the food requirements for a family with scarce capital and land. Diversity was the critical factor in using the scarce resources efficiently. Crops, animals and other farm resources were assembled in mixed and rotational design to optimize production efficiency, nutrient cycling and pest control. There are now several such biodiverse farms. The activities that were started and other outcomes are described briefly below.

Participation

In all the activities there was an increase in the level of women's involvement above what was anticipated. The poorer and more needy women have benefited especially from the programme. The project started with 25 women, and by 2001 a total of 432 women had registered. It has now extended to the five surrounding communities of Akwaduo, Swedru, Apiankra, Nnuaso and Homebenase. Crystallizing the needs of women into practical applications provided the increase in active participation.

Woodlot establishment

As a first step toward providing the women with an enduring stake in the village economy, a plantation of teak (*Tectona grandis*) and *Cidrella* was started in 1993. The choice was that of the women themselves. Teak was planted to provide electrical poles for the village and income from sales of poles, building timber and also furniture making. The *Cidrella* species was planted mainly to meet fuelwood needs as it can be harvested within a few months of planting.

The woodlot became one of the major breakthroughs for the project. Increased availability of adequate electric poles and fuelwood established goodwill toward the project in the eyes of the villagers, winning their trust, and convincing them that woodlots on degraded land would yield significant economic benefits. Nearly 10 hectares have now been planted with trees. Branches of teak were pruned for use in building a poultry farm. A further benefit of the woodlot has been the saving in time previously spent in fodder collection (Oduro, 1993). Formerly the women had to spend about four or five hours to cut and collect one headload (approximately 30 kg) of fodder. The same quantity can be collected from the woodlot in two hours.

Nursery establishment

In 1993 groups of four to ten women headed by a leader organized themselves to produce seedlings. The women provided the raw materials, including the site and seed, while the polythene bags were provided by the project. In the first year, the PLEC women's association raised nearly 20000 plants and these were eventually used in their agroforestry and watershed programmes. The number of seedlings produced increased over 80 per cent from 1996 to 2001. The nursery project has led to a sustained increase in income through sales of seedlings and has reduced the overall economic dependence on farm produce. It has also brought about a change in the attitude of the women by instilling a sense of ownership.

Agroforestry

Traditionally, fuelwood needs have been met from farmlands. One of the major reasons for resource degradation has been the increased consumption of fuelwood with growing population. It was essential for the project not only to evolve strategies to increase fuelwood availability but also to reduce its consumption in order to protect the plantations. Some studies strongly indicated that agroforestry on farmlands was a viable solution to the problems of fuelwood shortage (Oduro, 1993, 1994, 1995).

One of the secrets of the project's success has been the excellent interaction between the scientists and the women, as demonstrated in the agroforestry activities. Participatory techniques were used to prepare detailed microplans. The species to be planted were identified during the exercise and the preferences given were widely different. Women consistently gave higher ranking to fuelwood, timber and fruit species that give high cash yields and fodder. They gave very high ranking to *Margaritaria discoidea* (pepia) and *Celtis mildbraedii* (esa). In addition, *Dacryodes klaineana* (achia), *Ficus exasperata* (nyakeni), *Funtumia elastica* (fruntum), *Albizia zygia* (okro), *Rauvolfia vomitoria* (kakapenpen) and *Hevea brasiliensis* (para rubber) were also listed by the women as they provide fuelwood throughout the year (Oduro, 1997).

Savings accounts for women

Experience elsewhere has shown that increased women's income did not necessarily provide better development opportunities due to the constraints of the existing social structure and cultural practices within the village and family. Women traditionally have no control over their income. To avoid such problems and to provide the women with experience in understanding the operation of national banks and post offices, they were encouraged to open small savings accounts in a rural bank (Bosomtwi Rural Bank).

The women showed that they have a tremendous capacity to mobilize resources if assisted appropriately. The PLEC women's association used their savings to purchase planting materials, labour, animals for rearing, and nursery materials for raising seedlings for their farms to enhance agrobiodiversity. The activity also had an impact on the status of women in the community, who could talk proudly of their own group-money. This gave them self-confidence and assisted their overall empowerment.

Inclusion of women farmers in workshops

Women were trained in plant propagation techniques such as budding and grafting so that they could raise horticultural and ornamental plants on their own. This has made their nursery economically viable. Training in pruning enabled them to make interim use of fuelwood from the rehabilitated areas used as woodlots. Training techniques have not been instructive only; they have also included singing, drama and role-play.

Women leaders attended regional and international workshops with scientists and policy makers. There have been exchange visits to and from demonstration sites in other parts of Ghana. All this has had a positive impact on the status of women farmers in the community and has earned them greater respect.

Involvement in watershed design and management

Women designed watershed restoration and management in partnership with the scientists. In particular, they set about renovating a community pond (*denkyemni*) to restore aquatic communities, increase biodiversity and begin fish production. The women have inspired the community and together they constructed drainage to the pond using community resources.

Rare food and medicinal farm establishment

Women have established rare food crops and a medicinal arboretum. They know the remedies for stomach ache, broken bones, childhood diseases, traumatic injury, and women's problems. Together with the local community, threatened species have been collected for protection, propagation and transplanting. Scientists assisted them to label the plants, document their uses and establish the arboretum.

Income-generating activities

Scientists assisted women with improving local chicken, goat, sheep and duck production. The participatory approach was used when introducing different production and management techniques. The women have also been given training in improved methods of snail and mushroom production, bee keeping and processing of honey and beeswax. The new technologies enhance agrobiodiversity, increase the farmers' income and reduce pressure on the wild species. Responding to demand for a widening of income sources, adding value to indigenous leafy vegetables and cassava have been promoted, including bakery and pastry products from cassava flour.

Evaluation and experience gained

In the short span of seven years, the project has made tremendous progress in the empowerment of women. Women now participate in village meetings, talk of their work and needs, and openly seek recognition and visibility. The project has attracted the attention of visitors and the media both at the national and international levels. Women who had hitherto lived secluded lives, largely untouched by mainstream development, became excited about their new opportunities and wanted to make the most of them. The Unit committee elections of December 1998 brought about political changes in the village leadership and, for the first time, a woman, the leader of the PLEC women farmers association (Ms Osei), was democratically elected as a member of the Unit committee. However, the success of their work has brought its own problems. Some men in the villages still feel jealous and threatened by the women's empowerment.

Future scope and challenges

It takes a long time for village institutions to take full responsibility for managing biological resources in a sustainable manner. During the coming years, the project will continue to enlist greater support from women and their families with the aim of increasing their incomes so that more of their immediate family needs can be satisfied. Toward this end, the project has increased its emphasis on promoting income-generating activities that will help in achieving long-term sustainability of biodiversity conservation. Community development work on a long-term basis cannot be solely an operational activity of government because of the very nature of the work. What has been achieved with the PLEC project at Jachie has been unique, but if these processes are to be sustained there must be a planned and effective strategy. The project strategy and option for the future, in the absence of any effective NGO in the region, is to work towards promoting community-based structures. It gives tacit recognition that these farmlands belong to the women farmers and hence will be best managed by them. Along with the strengthening of Unit committees, farmers around Kumasi and in southeast Ghana have begun to urge the creation of PLEC farmer associations at village level throughout the country. A national or regional association could be formed with its own organization, which would be responsible for programmes using the PLEC approach. Self-help groups could be promoted, starting with savings and credit programmes and, over a period of time, these could be registered as a Rural Women's Bank. Fortunately in Ghana a Ministry for Women's Affairs has been established for the first time, to handle issues pertaining to women.

Thus, from simple actions and natural abilities, a truly emancipated group of rural women can play a vital part in national development. The most important steps taken towards achieving success were the establishment of the PLEC women farmers' association, and the recognition of the women farmers as important stakeholders. What are also needed are ties with similar women's groups elsewhere to share experiences and strengthen the project. Many such activities may seem outside the immediate purview of project activities, but holistic action can go a long way to ensuring achievement of PLEC's main objective – the sustainable management of biodiversity in Ghana.

Part III. Agrodiversity case studies

14 Agrodiversity assessment and analysis in diverse and dynamic small-scale farms in Arumeru, Arusha, Tanzania

FIDELIS B. S. KAIHURA, PAUL NDONDI

AND EDWARD KEMIKIMBA

AGRODIVERSITY, the diversity of cropping systems, crop species and farm management practices, is used as a way of spreading risk and supporting food security in resource-poor farming systems (Tengberg et al., 1998). Many of the smallholder farming communities in the tropics utilize the diversity of their environments, manage a large variety of crops and genotypes and apply a wealth of techniques both to exploit the diversity and support rural livelihoods (Richards, 1985; Pretty, 1995; Reij et al., 1996). Agrodiversity is a response of resource-poor farmers to inherent environmental variability, and is high in the semi-arid areas with characteristic temporal and spatial variability. Temporal variability is primarily a function of rainfall, while spatial variability reflects variation in the landscape, relief and soil type as well as in rainfall. Ethnic, cultural and economic diversity are superimposed, with differences in wealth and access to resources. Responding to such variability, farmers choose different land management strategies according to their asset holdings (Scoones, 1996), status and livelihood requirements. Their choices are revealed in an often-surprising heterogeneity of land uses, crops and practices, even in apparently homogeneous areas. The assessment conducted in this study was aimed at identifying salient practices and behaviours of small-scale farmers in the diverse and dynamic environments of Arumeru, Tanzania.

Assessment methodology

Assessment of agrodiversity was conducted in 1999–2000 at two sites in the Arumeru district, covering in detail most aspects of biophysical diversity, and crop, land and livestock management. Both sites are on the slopes of Mt Meru. The Olgilai/Ng'iresi site is at high altitude with relatively high rainfall, and is densely populated (2158 people), with only 0.22 ha of cultivable land per person (Murnaghan, 1999). Kiserian is at lower altitude, is semi-arid and, although more populous than Olgilai/Ng'iresi (3330 people), has more cattle than people. Both areas are occupied mainly by Waarusha people, with smaller numbers of Wameru, Wapare, and some Wachaga from the nearby Kilimanjaro region.

The first task of the study was to establish current Land-use Stages and Field Types using the methodology proposed in Chapters 5 and 6. For each Land-use Stage, reconnaissance traversing was done, crosscutting the entire area. In farmers' fields, current Field Types were identified and named. In all cases, identification and assessment was done in collaboration with key informants. The checklist outlined in Chapter 5 was the main guideline used for most of the data collected. In each Field Type, farm ownership, farmer category, location (geo-referenced), land form, vegetation, drainage and percentage slope of the field, fertility rating, and evidence of N, P and K deficiency symptoms on plants were recorded.[1]

Management systems were assessed in terms of crops and cropping systems, planting, tillage, livestock management, soil management of household farms and soil management of rented and hired farms. The types of crops and cropping systems found in the Field Type were recorded noting the scientific, Kiswahili and local names, the total number of species and varieties per Field Type, economic uses, characteristics of each variety, and the management strategies. Planting season and time, planting materials and methods, source of planting materials and presence of volunteer crops in each Field Type were recorded. Tillage was assessed in terms of types of tillage and tillage tools. Livestock were described by recording breed, source and purpose of the breed, and feeding and housing systems. Crop management practices such as types of weeds and pests and control, and crop storage were recorded. Fertility management strategies, soil erosion, drainage and moisture conservation systems were used to assess soil management.

An initial step was to identify volunteer farmers to work with. These farmers were grouped into three resource-endowment categories based on criteria set by farmers themselves. The criteria included the number of wives, the type of house (brick walls with iron roof versus mud with grass roof), the number and type of livestock, size of the farm and types of farm implements used (tractor, ox-drawn or hand). The majority of farmers in Olgilai/Ng'iresi were rich (63 per cent) while the majority in Kiserian were average (54 per cent). At both sites poor farmers were in the minority. Understanding farmer differences in resource endowment was a useful criterion to explain observed differences in resource management and other elements of agrodiversity.

Land-use Stages and Field Types in the study area

The five dominant Land-use Stages identified in the Olgilai/Ng'iresi sub-humid site were natural forest, planted forest (taungya system), agroforest,

Table 14.1 Description of Land-use Stages and Field Types at Olgilai/Ng'iresi, Tanzania

Land-use Stage	Field Types	Field type description
Natural forest	Least disturbed	Upper footslopes of Mt Meru; inaccessible due to steepness and deep incised valleys. Slopes 85–50%; humid tropical climate; some wild animals; area gazetted
	Slightly disturbed	Upper footslopes of Mt Meru; used for timber, firewood, and medicinal plants; distance from village and steepness limits use. Slopes 15–35%; humid tropical climate with few wild animals; area gazetted
	Highly disturbed	Cone-shaped hilltops sometimes used for recreation; used for timber, firewood, and medicinal plants. Tree harvesting controlled by village but most economic trees and shrubs already harvested
Planted Forest	*Pinus* with temporary cropping	*Pinus* trees planted after clearing natural forest; maize and beans commonly in rotation with cabbage and potatoes; crop combinations and sequences differ between farmers and seasons. Slopes 10–20%
	Cypress with temporary cropping	Cypress trees planted; cropping system similar to *Pinus* plantations.
	Eucalyptus plantation	Natural forest cleared and planted with eucalyptus only.
Agroforestry	Crops and trees	Complex mixes of crops and trees depending on farm size, season and farmer preference; coffee, banana and trees with maize and beans most typical. Varying slopes
	Maize and beans with trees	Maize and beans as intercrops with trees as hedges on contours and boundaries; the most economic crop(s) occupy the largest area
	Potatoes in rotation with vegetables	Commercial potatoes in first season, followed by cabbage and fallow in the third season of the year
	Maize	Maize planted as monocrop
	Potatoes	Potatoes as a commercial monocrop
	Farm boundaries	Boundary fences and partitioning structures with trees, shrubs and climbers. Species have diverse uses but most have thorns to restrict trespass
	Plot boundaries	Structures separating Field Types within farm, including crop residue and weed piles along boundaries, creepers and shrubs of economic value. These may be destroyed and spread for soil fertility improvement
	House gardens	Near the house with local and introduced vegetables. Mostly on flat areas or gentle slopes with irrigation
Water source	Micro-catchment	Delineated patches less than 30 m² protecting water seepage points; planted with perennial trees and bananas. No tree harvesting and trespass limited to fetching water; owned communally
Fallows	Regenerating fallows	Communal or individual plots temporarily left uncultivated for fertility recovery. Steep to moderately steep slopes
	Pastures, recreation or fallows	Lands left fallow and/or family recreation places; goats may graze
	Tethering and cut-and-carry fields	Pastures where cows are tethered for grazing or servicing (in case of bulls); grass may also be cut for fodder

water source micro-catchments and pasture fallows. Table 14.1 describes the identified Land use Stages and Field Types in May and June 1999. Agroforest was the dominant Land-use Stage, covering 80 per cent of the site. The different Field Types in each Land-use Stage were also identified. Most Field Types were on farmers' fields while some were in the forest (gazetted forest) and in open common grazing lands. A total of 42 different Field Types were found in Olgilai/Ng'iresi, with nine on one farm in Ng'iresi. The boundary and house garden Field Types had the greatest frequency, reflecting the importance of demarcations of land under conditions of land scarcity. House gardens were an important immediate source of household vegetables in the absence of cash. The most common vegetables grown were cabbage, onions, tomatoes, spinach, *Amaranthus*, eggplants and peppers.

In the semi-arid site of Kiserian the seven Land-use Stages identified were *mbuga*[2], mixed cropping fields, neglected fallows, woodlots, agroforests, stone-dominated hilltops and quarries (Table 14.2). *Mbuga* was dominant and covered 68 per cent of the area. There were 29 Field Types, and maize and bean intercrops and natural pasture were the most frequent. These common Field Types in Kiserian were consistent with the site being an agro-pastoral area.

Overall there was greater diversity of crops and cropping systems in the sub-humid densely populated area of Olgilai/Ng'iresi than in the semi-arid site at Kiserian. Individual Field Types were, however, bigger in Kiserian. In general, Field Types in individual farmer fields change with seasons. The greatest changes were at the sub-humid site, where there are three seasons a year and many diverse crops and cropping systems.

Soils and soil fertility

The soils of Arumeru are described in Table 14.3. Classification is according to farmers' ways of describing soils but the FAO/UNESCO classification is included to enable others to understand the soil types. The major criteria used by farmers for classifying soils were soil colour, workability, susceptibility to wind and water erosion, and fertility with reference to crop yield and water-holding capacity.

There were also minor soils recognized locally. In scientific terms, soil characterization was based on physiography and land use. There were 25 dominant soil types. Fertility rating ranged from high at the high-altitude, high-rainfall site (Olgilai/Ng'iresi), particularly in the planted forests and coffee and banana agroforestry systems, to low in the fallow and pasture plots at the semi-arid, low potential site (Kiserian). Based on visual assessment of the crops growing in the respective Land-use Stage and Field Types at the time of the study, most crops, trees and shrubs showed signs of nitrogen, phosphorus and potassium deficiency.

Table 14.2   Description of Land-use Stages and Field Types at Kiserian, Tanzania

Land use Stage	Field Types	Field type description
Mbuga	Overgrazed land	Lowlands and valley bottoms; surface cover less than 20%; deep gullies; cattle graze freely except where gullies are dangerous.
	Averagely grazed land	Pasture land with less than 40% surface cover in toe slopes and valley bottoms, sheet and rill erosion, pockets of depositional materials
Fallows	Neglected fallow	Undulating plains; seasonally cultivated but grain and straw transported to homesteads of farmers living elsewhere; declining in productivity
	Slaughter areas	Semi-permanent plots for slaughtering on large farms with many cattle
	Graveyards	Individual or communal areas for burial of relatives or clan members
	Toilet areas	Bushy plots away from the homestead; may later be used for grazing heifers. It is a cultural practice that parents, especially males, do not share toilets with children (children should not know that parents also go to the toilet until they reach maturity); household head has a separate area
Mixed cropping	Maize and bean intercrop	Typical cropping system for most farmers
	Maize and beans with trees	As above but with trees planted as a border on contours or edges of plots; tree species vary with farmer
	Maize, pigeon pea with trees	Pigeon peas replacing beans
	Chickpeas	Planted after the main crop to make use of residual moisture in flat lands in valley bottoms with heavy black clay soils
	Maize monocrop	Cropping system typical of resource-poor farmers
Agroforest	Mango-based hedges	Combination of mango trees with other trees and shrubs forming the farm boundary practiced by very few old farmers
	House garden	A combination of traditional and introduced vegetables planted near the house; mostly on flat parts of the farm; usually smaller than any other Field Types
	Fallow	Separated part of the farmland for tethering a few animals, and with scattered trees preferred
	Crop with banana and medicinal plants	Common crops and endangered species planted for various purposes
	Hedges/fences	Boundaries between farms and fields, used as cattle tracks or sometimes as cut-off drains and often planted with medicinal plants
Woodlots	Woodlots	Planted or conserved individual, communal and traditional forest areas. Individual woodlots get more attention and have a greater diversity
Stony hilltops	Extremely degraded lands	Grazing lands with exposed rock outcrops and most soil eroded, few shrubs and scattered grass patches. The area may be left as fallow
Quarries	Quarries	Completely degraded land with more than 80% rock outcrops; rock is moulded to make stone bricks or gravel. Characteristic in stream beds, valleys, hill tops and all places excessively grazed or cultivated

Table 14.3 Dominant soil types in Arumeru and their characteristics

Local soil name	FAO/ UNESCO	Fertility rating	Indicators	Remarks
Engulukoni nanana/nabuyavi	Eutric Andosol	Very high	Very deep with 15–20 cm decomposed surface litter, very dark brown to very dark greyish brown clay to silt clay	Friable; used for raising seedlings in nurseries
Engulukoni narok	Haplic Andosol	Low	Deep, dark brown clays with weak structure; underlain by strongly weathered basaltic rock; found on summits and steep slopes	Loss of topsoil due to erosion, low fertility, moderate to heavy administration of inputs required for good production
Engulukoni nador	Haplic Luvisol	Low	Weakly structured dark reddish-brown clays; found on upper and middle slopes	Traditional soil fertility measures necessary or use of chemical fertilizers
Engulukoni nador	Haplic Nitisol	Moderate	Deep red clays with shiny faces	Evidence of moderate to severe wind and water erosion on farmlands and grazing lands; continued cultivation without inputs and other conservation measures reduces fertility
Engulukoni sero	Vertic Luvisol	Moderate	Intergrade between reddish-brown clays and heavy deep black clays; cracking when dry; salty taste in subsurface; found on lower slopes	Production improved by additional nitrogen and erosion control; timelines in ploughing is crucial as soils become sticky and plastic when wet
Engusero	Calcic Vertisol	Moderate	Dominance of *Acacia* species, imperfectly to poorly drained heavy cracking clays, with many subsurface $CaCO_3$ concretions and manganese nodules	Drainage structures improve production; overgrazing contributes to observed erosion
Oldenderit nador		Moderate	Deep red clays with shiny faces or slickensides; found on summits	Neglected soils prone to water and wind erosion with exposed rock outcrops; a large area cropped and both grain and stover transferred to uplands or sold
Rigirigi Regosol	Haplic	Very low	Very shallow soils	Land that is used to collect murram for roads and house construction, making of stone bricks

Note: The Indicators were developed by both farmers and scientists. The major types of soils and their characteristics by farmers were all reflected in the scientific classification.

Woodlots

Conservation of remnant natural forest patches was an important activity of some farmers, particularly elderly farmers. Different types of trees and shrubs were found in complex mixtures, each of which had a known economic value. Most remnant forest woodlots were found in ravines and in old or seasonal river valleys. Such woodlots were conserved as gene banks for plants used for traditional medicine, building poles, firewood, soil fertility improvement, and wood for making carvings. Other economic uses of the trees included fruit production, dyeing materials, shade, wind-breaks and drought insurance. Farmers understood the growth habits of each type of tree and managed them differently. Trees least attacked by pests and diseases received less attention than more vulnerable types. Although in Kiserian there were communal woodlots that were less carefully managed, most woodlots were privately owned just like the crop fields. They were normally harvested once a year.

Pastures

Pastures constituted a Land-use Stage or Field Type according to the way they were used. *Mbuga*, typical in Kiserian, was mainly for grazing, and livestock grazed in extended areas of natural pastures. In Olgilai/Ng'iresi, pastures constituted a Field Type in the cropping systems. They were natural, or planted for tethering or cut-and-carry fodder. Some farmers without livestock had pasture grass for sale. Pastures for fodder were also planted to strengthen conservation structures like contour bunds. Such fodder was cut and sold or given to livestock at home. Some grass was also used for roofing, and provision of soil surface cover and organic matter. Most farmers mixed grasses with legumes to improve nutritional quality and palatability, and to improve soil fertility through nitrogen fixation. Both woodlots and pastures were mainly found as complex mixtures of different species, either natural or planted.

Crops

The major crops included maize, beans, banana, coffee, potatoes and sweet potatoes. Most were grown as mixed plantings, together with grass and vegetables. Table 14.4 details the varieties found in Arumeru of the three major food crops. In both sites the agroforestry Field Types had the greatest diversity of crops and varieties. In some farms more than five varieties of a crop were found in the same field.

Planting time, methods and materials

At Kiserian, following the usual practice, maize was planted just before the beginning of the rainy season, which starts between mid-February and early

Table 14.4 Varieties of the main food crops in Arumeru

Crop varieties	Plant characteristics and crop uses
Maize varieties	All are used for food, income, and crop residues fed to animals
Kienyeji	Not very sweet, low yielding, drought susceptible, tolerant to storage pests, good milling quality
Katumani	Drought tolerant, early maturing, low yielding, good milling quality, tolerant of storage pests
CG4141 (Lowlands)	Good milling quality, drought tolerant
UCA (Highlands)	Good milling quality, drought tolerant
Kilima	High yielding, high water demand, susceptible to storage pests, good milling quality, high-quality flour
***Phaseolus* varieties**	All are used as food, most residues are fed to livestock, and some are used for income
Soya kijivu	Income, high price, good taste, 'no gases after eating', early maturing, sweet, grey climbing type
Kachina	Income, high price, early maturing, spoils quickly after cooking
Lovirondo	Climbing type, 'gases after eating', laborious to harvest, low market price
Bwanashamba	Most popular in Kiserian, high yielding, good taste, susceptible to diseases and aphids
Maasai red ndogo wide(namira)	High yielding, good taste, 'no gases after eating', needs spacing for high production
Maasai red kubwa (namriri)	Income, high price, early maturing, bush type, susceptible to diseases, good flavour
Karanga	High yielding, good flavour
Lyamungu 90	Income, high price, good flavour, early maturing, drought tolerant, high yielding,
Kiburu	Drought tolerant, grows well on soils with poor fertility
Engichumba	Income, very high yielding, violet bean
Engichumba-ng'iro	High yielding, sweet, grey bean used in *loshoro**
Engichumba-narok	Income, similar to Engichumba-ng'iro, black bean
Moshi	Income, very high yielding, very sweet, yellow bean
Kibumulu	Income, fast cooking, high price, dark red bean
Banana varieties	
Kisimiti	Early maturing, drought tolerant, used for income, brewing, fodder (stem)
Ng'ombe	Hard when cooked; used for *loshoro**, brewing, income, roofing, fodder
Mshale	Long and thick fingers; used for roasting, *matendela**, income
Uganda fupi	Early maturing, susceptible to pests and diseases; used for soup (*mtori*), fruit, income, fodder (peels)
Uganda ndefu	Large with few fingers, susceptible to pests and disease; used for soup, fruit, fodder (peels)
Kisukari	Very sweet, drought and disease tolerant, low nutrient demand; used for fruit, income, fodder (stems)
Mzuzu	Tolerant to drought and disease; roasted for evening meal
Malindi	Drought tolerant; used for food (*matendela*), fodder
Mnanambu	Used for shade, soup, roasting
Mkonosi	Disease tolerant; roasted
Mkono wa tembo	Disease tolerant; roasted
Ndishi	Susceptible to diseases; used for *loshoro**, income
Olmuririko	Modest tolerance to diseases; used for *loshoro**, brewing

*Loshoro: traditional Waarusha/Wameru/Wamasai food made of maize, beans and milk.
*Matendela: traditional Waarusha/Wameru food made of vegetables, milk, banana and beans.

March. Beans were planted at the same time, although a few farmers planted beans after maize germination. During the middle of the season, sweet potato cuttings were planted, but mainly in outlying fields. This later planting was to avoid tuber rot caused by heavy rains. Other crops planted were onions, a local *Amaranthus* species grown as a vegetable, and sisal planted from its bulbils as boundary vegetation.

Maize was sown in rows and intercropped with beans, the latter being generally two or three times more numerous than maize rows. Beans were planted more densely because they stand a greater chance of survival if rains are inadequate, and fetch a much higher price at local and commercial markets. In some farms pigeon peas or cowpeas were broadcast before the maize and beans were planted in rows. Seeds of local varieties were obtained from farmers' previous crops, while improved seeds were obtained from commercial farmers, and the Zonal Agricultural Research Institute. Vegetable seedlings were grown in nurseries after buying seed from farmers' association shops, retail shops or agricultural institutions. Farmers also propagated wild amaranth seedlings. Other planting materials obtained from farmers' own farms included fodder, sweet potatoes and sisal.

There were very many crops and crop combinations at the high-altitude, high-rainfall site of Olgilai/Ng'iresi. The long rains start in April, and the crops grown in that season include potatoes, tomatoes, coffee and fodder (both grass and trees). Selected forest trees were also planted. The crops that were planted during the short rains that start in July or August included maize, bananas, sugarcane, beans and sweet potatoes. Vegetables (e.g. cabbage) and fodder were planted during both seasons. Boundary trees and shrubs, and yams were also planted throughout the year.

Most vegetables were planted using seedlings raised in their own or commercial nurseries. Several farmers bought seed for nursery preparation. Potato growers used seed tubers from the previous crop or bought from the local markets. Coffee seedlings and banana suckers were obtained from the farmers' own nurseries or farms. Coffee seedlings were sometimes obtained from commercial nurseries.

Seedlings for the planted forests were obtained from the forestry unit of the district or region. Others obtained their planting materials from neighbours, friends, private shops and farmer organizations. Although farmers were encouraged by expert farmers to plant improved varieties, it always takes some time before new seed is adopted. Local seeds were always planted because they are sure crops to harvest in case of bad weather.

Tillage management

In this area of management there were major differences between the two communities. At cattle-rich Kiserian with its larger farms, primary tillage for most annual crops was by ox-ploughing. This was normally preceded by spot application of manure. This was spread as the oxen were drawn so that it was incorporated together with crop residues. Weeds were overturned at the same time. Secondary tillage was re-ploughing using the same tools to break the clods and make furrows for planting. In areas where the farms were along steep slopes, soil clods were left to reduce water erosion. Pre-application of farmyard manure was also practised on vegetable farms. For planting grasses such as *Desmodium* and *Chloris gayana*, the clods left after primary tillage were smoothed and a fine seedbed was made using hand hoes before the seeds were broadcast and covered. For farmers without oxen, primary and secondary tillage were done by hand hoes. Hoes were also used on small plots and on soil conservation structures; but many farmers with oxen did not even possess a hand hoe.

Farms in Olgilai/Ng'iresi were smaller and local cattle were few. Most farms were on steep slopes and ox-ploughing was almost non-existent. Primary tillage for most annual crops was done using forked hoes, and also included the incorporation of crop residues and weeds from the previous season. Secondary tillage was not normally done as seedbed preparation was completed during primary tillage. Slashers were used on farmers' fields mainly in Kiserian. For potatoes, tomatoes, and *Amaranthus*, forked hoes or hand hoes were used. In planted forests, machetes and axes were used to clear bushes, while hand hoes were used for cultivation. Similarly slashers and forked hoes or hand hoes were used in management of coffee and banana farms, where the use of mulching systems avoided the need for tillage except at planting.

Soil fertility and water management

There were various soil management practices, but traditional practices were most common at both sites. At Kiserian, physical structures were few and were mostly reinforced with trees and grass materials. They included contour ridges in maize/bean cropping systems for soil erosion control, deep trenches 0.2 m deep × 0.7 m wide and covering 100 m for soil moisture conservation in natural pastures, and shift grazing in improved and fenced pastures. Stone-lines were used to control soil erosion in woodlots and maize fields, and deep tillage was used for soil moisture conservation in maize and bean cropping systems. In *mbuga* pastures the prohibition of cultivation and burning was important for erosion control.

At Olgilai/Ng'iresi there was a similar mixture of methods, but there was greater use of chemical fertilizers by the better-off farmers, especially for potatoes, cabbage, maize and beans, sugarcane, and bananas. Physical structures were not common, but included fences and raised beds along boundaries for soil erosion control, *fanya juu* (thrown-up terraces) for maize and sweet potatoes, and cut-off ditches in pasture and maize and bean plots. Ditches controlled both erosion and drainage. *Fanya chini* (narrow, cut-down terraces) were usually 0.3–0.6 m deep and 1 m wide, with lengths that varied greatly depending on field size.

Soil fertility was maintained by incorporating crop residues, house refuse, weeds, manure and ash. Crop remains were also fed to livestock and the manure produced on tethering pastures was collected and applied to fields. The remaining part was directly used to fertilize the fallow. *Grevillea* and *Sesbania sesban* were planted as soil improvers and their litter was used in various Field Types. In the coffee and banana agroforests, weeds were collected together and left to decompose before incorporation. Materials difficult to decompose were mulched under the coffee, bananas or other trees.

Trashlines were the most common traditional erosion control practice. In most cases lines were made across the slopes and sometimes on contour ridges. Crop residues and stover, and other plant residues were traditionally used to conserve soil moisture. Canopy cover in coffee and banana, and grass or sweet potato ground covers were used to conserve moisture and control erosion. Natural vegetation on contour ridges was enhanced by planting elephant grass (*Pennisetum purpureum*), sugarcane, fodder grasses, *Cypress*, bananas, *Leucaena*, *Sesbania*, *Grevillea*, *Seteria* and other species. *Seteria* was also used for erosion control on boundaries and as live fences in vegetable gardens.

Sunken beds around coffee trees had a dual purpose. They collected surface litter for soil fertility improvement; and provided *in-situ* moisture conservation and soil erosion control by reducing surface runoff. In the sunken beds manure and compost were quite heavily applied.

There were also irrigation channels, using water from forest catchment streams, for irrigation of potatoes, maize and beans, and house-garden vegetables in some farms. The irrigation channel at Ng'iresi was mainly used for household water supply, with only some watering of horticultural cash crops, but irrigation was used for all kinds of crops at Kiserian (Murnaghan, 1999). In the larger Nduruma basin to the east there was an elaborate management system with committees for each furrow, careful and well-monitored regulation of supply, and penalties for infringement of quite precise by-laws. There was also careful local regulation at Kiserian and Olgilai/Ng'iresi. Water Use Associations managed the use of the indispensable water resources, especially through the dry season.

There were local regulations in both communities for the protection of forests and woodlots. In conserved forests, these included prohibition of free cutting of trees, burning, collecting firewood and cultivation. For planted forests, firebreaks of bare land were also made. The planted *Pinus* forest in Ng'iresi was protected by firebreaks 6.5 m wide and more than 1 km long.

Weed and pest control

Occurrence of different types of weeds depended on landform, soil type and management regime. Different weed populations characterized the two sites. At Kiserian, weeding was done at least twice during the growing season, using forked hoes. Other practices included slashing or manually uprooting. Some weeds such as couch grass were burnt after weeding, and some were fed to livestock. Weeds were controlled in all Field Types including pastures, planted forest and agroforestry areas

There was great diversity among farmers in pest and disease management. Chemical spraying was common, but a wide variety of other methods were also used. Pesticides were mainly used to control pest attacks rather than for prevention. The cultural methods used included early planting, pruning, planting many seeds per hole to reduce damage by maize stalk borers, and changing planting materials controlled the weevils that are common in sweet potatoes. Manual methods were used to control other pests; seasonal grasshoppers, cut worms, stalk borers and aphids were reduced by hand picking. Coffee berry borers were wire-hooked and weevils in banana were controlled with hot water or by destroying infected shoots. In agroforests, predators including birds and chameleons reduced pest numbers. There was no evidence of serious pest attacks in pastures. Some farmers kept cats or used rat poison and traps to reduce numbers of rats and mice in the fields.

Traditional mixes like pepper plus tobacco and kerosene mixtures were used in conjunction with chemical sprays to control white stem borers in coffee. A urine, pepper and onion mix (*kitunguu swaumu*) was used to control aphids in the coffee and banana agroforest. For vegetable producers, cattle urine was used to control caterpillars, cutworms, aphids and *siafu wekundu* (brown ants). Ash was also used in various ways. Some farmers used it to control stalk borers. Application of ashes was also known to control root rot in potatoes and cabbage. In house gardens it was used to control *siafu wekundu*.

Storage of maize, coffee and beans was important. Most other crops were quickly marketed. Scania, the large grain borer, is a serious introduced pest of stored maize. Bagged maize was usually treated with a chemical (Actellic), but airtight jars and external structures of traditional type were also used. Traditional control measures included the use of ash, dung, *magadi* (soda ash) and sun drying. Some farmers stored the grain in cold rooms, and others mixed in tobacco seeds or allowed chickens to feed. Seed maize was not hulled but dried and then hung above the cooking place or spread in the shade.

Livestock

Arumeru is a cattle-keeping district, but while there were many cattle at Kiserian, there were relatively few at Olgilai/Ng'iresi. Most Kiserian farm-

ers kept the local breed, the Tanzania Shorthorn Zebu, and a few introduced breeds for crossbreeding. Visited households had up to 40 cattle. The
major source of the improved breeds was the Heifer Project International,
while local breeds were either bought from local markets or inherited.
Livestock were kept to provide income through sales of milk, meat and
cattle, milk and meat for home consumption, manure, and draught oxen
(especially the local breeds). Other purposes included celebrations and
prestige. Most of the local breeds (cattle, sheep and goats) were grazed
freely and kept in open kraals, while improved breeds were not grazed and
were kept indoors. A few farmers kept their livestock in their own houses.
Livestock pests and diseases were often treated with traditional herbs and
mixtures.

Cattle breeds found in Olgilai/Ng'iresi included Friesian and Ayrshire and
their crosses, together with the local Tanzania Shorthorn Zebu. Goats were
mostly of the local types with a few hybrids. Sheep were mainly local. The
number of livestock ranged from zero to five, and most farmers bought them
from local markets. Others obtained them through dowry payments or inheritance. Both NGOs and religious groups provided improved bulls for
cross-breeding. Livestock were kept for the same purposes as at Kiserian.
Sheep were used for traditional sacrifices and prestige. Livestock at
Olgilai/Ng'iresi were either zero grazed, or more rarely tethered and grazed
in small fenced plots. All were kept indoors in separate sheds or, in a few
cases, in the farmer's house.

Management of bought, borrowed and rented support farms

Because of land scarcity most farmers in Arumeru had additional, distant
support farms away from home. Some of the farms were as near as 1 km
away, but some were as far as 100 km away. Discussions with farmers
indicated that most of them grew maize and beans in support farms. Crop
residues from rented or borrowed farms were transported home to feed
livestock, or were sold to farmers in need. Only on personal plots obtained
through purchase were some residues left and incorporated into the soil
for fertility and moisture improvement. Because of large herds of cattle for
milk and manure production in Kiserian, several farmers kept only the
improved breeds and local oxen for ploughing at home, and sent the rest of
the local free-grazing animals away to distant places, sometimes in a
different district.

The same practice was common for Olgilai/Ng'iresi farmers. Most farmers
had plots in the regulated taungya systems, where pines, cypress and eucalyptus trees were planted together with various crops. After tree planting,
farmers were allocated land for crops until there was 56 per cent or more
canopy cover and it became unprofitable to grow crops. Farmers then left the

land and trees under the care of the forest department. Others had additional farms outside the taungya system. In all cases, fertility was supported by *in-situ* decomposition of forest litter. While the assessment team did not inspect many of the outlying sites, it seems improbable that agriculture practices on them were sustainable.

15 Biodiversity as a product of smallholder responses to change in Amazonia

MIGUEL PINEDO-VASQUEZ, CHRISTINE PADOCH,
DAVID MCGRATH AND TEREZA XIMENES-PONTE

Background

THE GLOBAL BIODIVERSITY CRISIS is forcing countries to measure their biodiversity, assess the threats, and propose actions to conserve, restore or improve species conservation and ecosystem and landscape diversity (UNEP, 1995). Results of these biodiversity inventories are helping to identify regions and countries where species and ecosystem extinction are at critical levels (Vogt et al., 2000). Although extinction is a biological problem, the solutions are also dependent on social factors (Kalliola and Flores Paitan, 1998), and a thorough examination of the processes that change biodiversity levels over time and space is needed (Brondizio, 1996).

Research conducted in Amazonia has provided two distinct discourses on the processes that control biodiversity levels. Ecologists aver that biodiversity is the result of natural processes (Terborgh, 1999). In contrast, the anthropocentric point of view, proposed by many social scientists, argues that biodiversity in Amazonia is the result of long-term human intervention in and manipulation of natural processes (Raffles, 1998). Both arguments commonly demonstrate limited appreciation of the complexity and variability of interactions of natural and social processes that create high or low levels of biodiversity. This study adds to the growing body of research that describes how existing biodiversity is the result of complex, highly variable and contingent interactions of natural and social processes.

Measuring biodiversity in the landholdings of smallholders offers an opportunity to look beyond the environmental ideologies that cause many researchers to ignore land 'tarnished' by humans and to search for 'pristine' ecosystems. We examine the role of smallholders in the formation and transformation of biodiversity within the estuarine *várzea* floodplain of Brazilian Amazonia (Figure 15.1). We emphasize particularly how recent changes in markets for agricultural and forest products are leading to changes in the way local people manipulate vegetation and in the resultant biodiversity of their environments.

Several authors have reported that the landholdings of Amazonian smallholders contain high levels of species and ecosystem diversity (Pinedo-Vasquez, 1995; Brondizio and Siqueira, 1997 and others). Studies that focus on swidden-fallow or shifting cultivation practices have found that

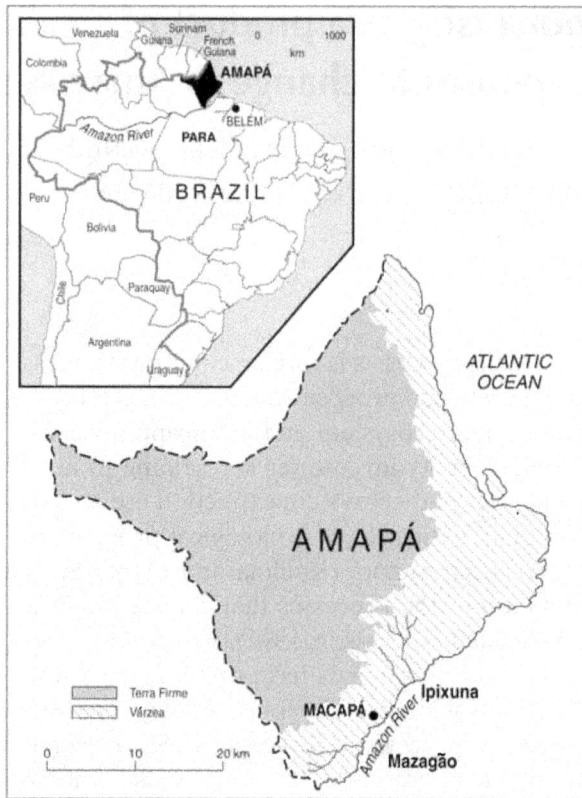

Figure 15.1   *The estuarine várzea floodplain region of Brazilian Amazonia*

agricultural fields, agroforests and forests managed by smallholders contain much more biodiversity than do more 'modern' land-use systems such as cattle ranches, industrial tree plantations and others that are part of the development models currently being promoted in Amazonia (Anderson and Ioris, 1992; Brondizio and Siqueira, 1997). We measured and monitored the levels of biodiversity produced, managed, or conserved by smallholders. These rural Amazonians take advantage of fluvio-dynamic and other natural processes to promote and conserve agrobiodiversity and biological diversity in their landholdings in *várzea* environments for their economic gain.

While most researchers agree on the ecological value of biodiversity, many argue that biological diversity does not help farmers improve their living conditions (Smith, 1996). In contrast, the conversion of forest into pastures and the subsequent reduction of biodiversity produces high economic returns for cattle ranchers (Fernando Rabelo, personal communication). The low economic value of biodiversity explains why in some tropical regions that are rich in biodiversity the majority of people are poor (Peluso, 1992). Similarly, some researchers argue that Amazonian smallholders have little appreciation for biodiversity (Smith, 1996). These points continue to be argued largely because the information that exists on managed biodiversity is incomplete, has yet to be presented in a consistent and accessible form, and most researchers do not understand the role that biodiversity plays in the livelihoods of farmers.

Most of the studies that measure species diversity managed by smallholders actually measure only a small fraction of the total diversity present. Some evaluate only house gardens and not the many other Land-use Stages or Field Types (Anderson and Ioris, 1992). Here we report on the existing biodiversity used and maintained in all the fields, fallows, house gardens and forests that make up the complete rural landholding of sampled smallholders located at the PLEC site at Mazagão-Ipixuna, Macapá (Figure 15.2).

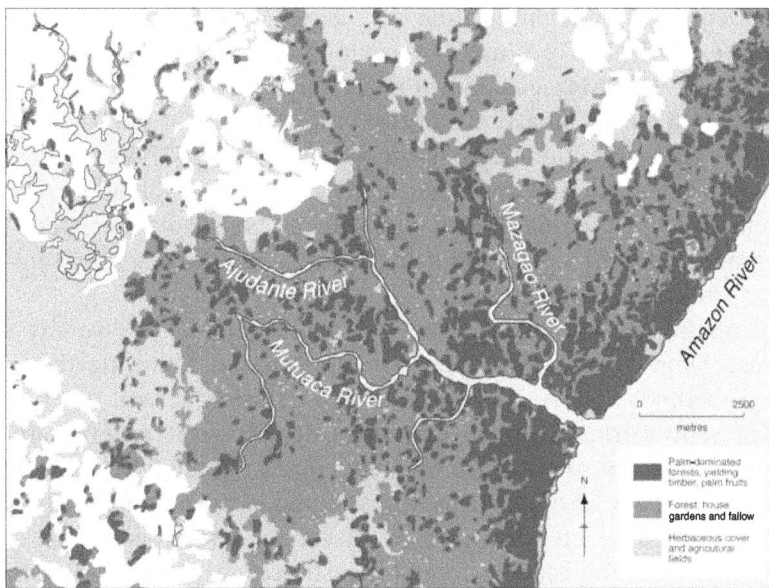

Figure 15.2   *The Mazagão demonstration site area*

The identification of smallholders as principal stewards of biodiversity raises new questions and requires a detailed understanding of the factors that influence and change species diversity. Researchers must avoid the tendency to reduce studies to the simple quantification of biodiversity. To observe existing biodiversity beyond a simple summation of numbers, we incorporated both social and ecological methods and techniques into the design of our surveys. We included ecological parameters to quantify biodiversity in different land uses using methods recommended by PLEC (Chapters 6 and 8). We combined these methods with participant observation and other techniques used by social scientists to identify the technologies and local knowledge that create and maintain biodiversity.

Methods

Pereira (1998) reported on physical changes that have affected the landscape of the site, including alterations in vegetation cover. Information was also gathered from aerial photos and Landsat images, as well as from land surveys, to identify important natural and social events that influenced the landscape in both sites. Data about the natural and anthropogenic origins of water bodies, land formations and type of vegetation cover were analysed by Raffles (1998), and used to help understand the role of people in modifying the landscape. Records on economic booms and busts, changes in land tenure and use over time were collected through interviews and by reviewing documents in archives located in Macapá and Belém.

The identification, quantification and classification of the existing Land-use Stages and Field Types used information collected from a sample of 60 landholdings; 36 in Mazagão and 24 in Ipixuna. Data from each landholding were gathered during an average of two visits per year, at the beginning and the end of the agricultural season, for the five years 1995–99. The number, area and location of Land-use Stages and Field Types in relation to the river and houses were recorded. With the help of household members (men, women and children) we drafted maps for each landholding representing its location within the landscape.

The existing agrobiodiversity and other forms of plant diversity in the Land-use Stages and Field Types were inventoried following the recommended methods and design. The indices used are discussed in Chapter 8, which presents worked examples employing data drawn largely from this site. Plant diversity in forest areas was measured in ten randomly selected samples (five from each site) from 27 ha of forest found in the 60 selected landholdings. All selected forests were managed at varying intensities for the harvest of timber and several other marketable resources. All trees greater than or equal to 5 cm diameter at breast height (DBH) were inventoried in four randomly selected plots of 25 × 25 m in each of the sample forests. Individuals with a DBH of less than 5 cm were inventoried in a sample of 80 (four in each 25 × 25 m plot) randomly selected subplots of 5 × 5 m. In each inventoried plot and subplot we recorded the common name, height, DBH and location of each individual. Because of the heterogeneity of the landscape we estimated β-diversity in addition to the species richness and Shannon Index.[1] Rank abundance per species was also calculated. Importance values for species and families were estimated for all sample forests. The majority of the results and the estimated floristic and structural parameters of the inventoried forests are included in the thesis of Fernando Rabelo, a student member of the Amazonia team (Rabelo, 2000).

Plant diversity in fallows was quantified in a randomly selected sample of 18 ha; 10 ha in Mazagão and 8 ha in Ipixuna, from a total 52 ha of fallows. The selected 18 ha sample included fallows ranging from five to eight years old that were managed for production of several marketable products, including the fruit of the açaí (*Euterpe oleraceae*) palm. Inventories were carried out in a total of 90 randomly selected plots (five in each hectare of fallow sampled) measuring 20 × 20 m. Natural regeneration was also inventoried in each fallow plot using subplots of 2.5 × 2.5 m. The same data were recorded for each plant inventoried, as in the forest plots. All biodiversity indices, including β-diversity, paralleled those calculated for the forest samples.

A sample of 36 house gardens (20 at Mazagão and 16 at Ipixuna) were also selected to measure plant diversity. The sampled house gardens covered a total area of 94 ha. The average size of each garden was 2.3 ha. Each plant in these gardens was inventoried. Data recorded included common name and growth habit (tree, shrubs, vine, grass and herb). Species diversity was estimated using the indices recommended (see Chapter 6).

We quantified species and varieties of planted or protected crops in ten fields in Mazagão and ten in Ipixuna, randomly selected from an average of 72 fields made by farmers at the beginning of the wet and dry seasons. Crop inventories were conducted twice a year before harvesting. The average size of the selected fields was 1000 m² (50 m long × 20 m wide). The fields were then divided into subplots of 10 × 10 m. Plant inventories were conducted in a sample of 100 randomly selected plots (five in each of the 20 sampled fields). For all planted or protected species and varieties found, we recorded common name, growth habit and uses. The levels of agrobiodiversity found in each sampled field, as well as variations from one season to the other, are discussed based on absolute and relative abundance.

Agrodiversity and other production and management technologies, as well as conservation practices, were recorded using participant observation techniques. Team members followed the members of the selected households in their daily activities and observed and recorded production and management techniques used in the fields, house gardens, fallows and forests. Information was cross-checked during group discussions and dialogue with the most knowledgeable members of each household and community.

Results

Dynamics of the natural environment

The Mazagão and Ipixuna sites are part of the estuarine *várzea* floodplain and are tidally influenced environments. Both sites are composed of very heterogeneous landscapes that include a great diversity of human settlements, land formations, waterbodies and vegetation cover. They are subject to complex and dynamic natural and anthropogenic processes. Although these processes are highly variable and unpredictable in their frequency, intensity and spatial characteristics, historical and landscape data show that the residents of both sites have developed technologies and strategies for managing and maintaining these processes.

Major changes in the direction of the Amazon River have taken place at both sites in the past 50 years. One of the major changes in landscape and river dynamics during this period was the formation of an island near of the city of Macapá. The residents of Mazagão reported an increase in sedimentation leading to the formation of new lands and an increase in the height of their levees. The residents of Ipixuna reported major changes in the size and number of streams, landforms and vegetation cover since the appearance of the island.

Our informants, supported by data collected from aerial photographs and Landsat images, record three main fluvio-dynamic events that have produced major changes in the social and natural landscape of Ipixuna. First, with the formation of the island, the river current became stronger and navigation increasingly dangerous. With the stronger current in the Amazon River, most of the sediments and other materials transported by the small neighbouring Pedreira River were deposited at the mouth of the river. As the current became stronger, any removal of vegetation along the stream also greatly increased lateral erosion by the tidal flow.

People from Ipixuna manage these fluvio-dynamic and other natural processes to open up new stream channels using water buffalo, in order to gain access to new land and resources (Raffles, 1998). They have developed intricate technologies for raising planting surfaces above the level of tidal floods (Padoch and Pinedo-Vasquez, 1999). These types of interventions exemplify how the mechanisms influencing landscape, ecosystem and species diversity cannot be identified by the study of either natural or social processes alone.

Social processes

Based on collected oral, historical and geographic information, the social processes characteristic of the two sites appear to be as complex and dynamic as the natural processes. The Ipixuna and Mazagão *várzea* floodplains have been subject to different intensities and degrees of land and resource use, and of outside intervention through the implementation of development and conservation projects. Resources have been commercially produced and extracted for varying periods encompassing several economic booms and busts. In Mazagão, commercial production of crops, such as rice, began in the middle of the seventeenth century. Commercial agricultural production in Pedreira-Ipixuna did not begin until the late 1940s (Raffles, 1998).

Despite decades of political and economic pressures to adopt modern production systems and reduce crop diversity, most of the production systems and techniques used by smallholders at both sites are based on locally developed technologies. A system of swiddening and management of fallows is the predominant production system practised. This allows smallholders to maintain high levels of agrobiodiversity and other forms of biological diversity in their fields, fallows, house gardens and forests. This in turn helps farmers gain cash income by harvesting or extracting many different resources.

Commercial extraction of timber has followed patterns similar to those of other areas of the estuarine *várzea*. Commercial logging started in both locations at the beginning of the twentieth century (Padoch and Pinedo-Vasquez, 1999). Extraction of seeds of murumuru (*Astrocaryum murumuru*), andiroba (*Carapa guianensis*) and pracaxi (*Parkia* spp.) was carried out commercially from 1940 until about 1970. Commercial fishing of catfish and shrimp became an important cash source for the inhabitants in both regions after 1950 (Raffles, 1998). Extraction of the fruits and the heart of the açaí palm has had a significant economic impact at both sites since 1960. Currently, heart of palm, fruits of açaí, several species of timber, and shrimp are the most important products extracted for sale.

Increases in market demand for açaí fruit and timber, and a decline in prices for agricultural products, have changed the economic roles of agriculture, agroforestry and forest product extraction. In the past, much crop production was destined for the market and was the main source of cash income for smallholders in both places; agroforestry and forest products were mainly used for subsistence. Currently, agricultural products are mainly used within the household, while agroforestry and forest management have become the main activities that provide income. This change explains why agricultural fields have become fewer at both sites.

Agricultural fields

Although quantitative data were collected for only two years, we can clearly see a trend in the number and size of the agricultural field Land-use Stages at both sites. In 1998, within the landholdings of Ipixuna and Mazagão households there were 25 and 23 fields, respectively. The 1999 data show that the number of fields declined to 12 and 18. In both cases fields were left to become fallows for the production of several marketable agroforestry and forest products such as açaí fruit and pau mulato (*Calycophyllum spruceanum*), a valuable fast-growing timber species.

While the number of fields declined, the average field size remained at 0.6 ha for both years. Results at both sites show that the decline in the number of fields did not affect the levels of agrobiodiversity in the fields. Farmers were not only planting crops in their fields, but also protecting the seedlings and saplings of several forest and agroforestry species. While a similar number of species and varieties were planted at both sites, Mazagão fields contained more than twice the number of protected species and varieties than found Ipixuna plots.

This variation reflected an important difference between the two sites. Farmers in Mazagão tended to focus on protecting rather than planting species; in Ipixuna the converse was true. Species and their varieties observed in fields included grains, tubers, fruits, medicinal and timber species. The average number of species and varieties of crops found in the sampled fields was higher than those reported in fields owned by smallholders within colonization projects and areas owned by cattle ranchers (Anderson and Ioris, 1992). A considerable number of the species inventoried were perennial.

The protection and planting of seedlings demonstrate how smallholders respond to the drop in prices and market instability of the typical annuals. Agricultural products produced in large industrial plantations in the south of Brazil are generally cheap and of high quality. As markets in urban Amazonia became flooded with staple crops from the south, smallholders found they could not compete with the falling prices of rice, beans and other crops. The farmers of Mazagão and Ipixuna responded by gradually shifting from a dependence on agricultural products to agroforestry and forest products for their cash incomes.

The increased importance of protection or planting of timber, fruit and other tree species in the fields also changed other farming technologies used by smallholders. Although fields were still made using swidden techniques, we observed that most farmers were opting not to burn the slashed material. They believed that seeds of valuable agroforestry species such as maracujá do mato (*Passiflora grandiflora*) and forest species such as tropical cedar are destroyed by fire, preventing natural regeneration. Similarly, weeding operations tended to be more selective and less intensive. For instance, when growing maize, beans and other annual crops, the majority of farmers did not weed their fields.

Fallows

Changes in farming operations and technologies were helping smallholders manage seedlings and saplings of tree and shrub species in their fields. In fields it was common to find protected species naturally regenerating both randomly and in clusters. Most farmers explained that they were protecting the seedlings of timber species, açaí palm and other valuable species to enrich their future fallows and forests. Based on the performance of the protected seedlings in the field stage, farmers developed two main fallow types: wild fallows (*capoeiras com mato*), where vegetation was low in valuable species; and enriched fallows (*capoeiras contaminadas*), where vegetation was dominated by valuable species. Smallholders subsequently used the wild fallows for making fields and managed the enriched fallows for the production of agroforestry and forest products.

Despite the assumption that human intervention in fallows lowers their species richness (Anderson and Ioris, 1992), we found that enriched fallows contained levels of plant diversity similar to those of wild fallows. We detected a general trend for smallholders to maintain or, in some cases, increase biodiversity in their plots as part of a strategy to increase the number of outputs available from the fallows. Of the eight sampled households, six maintained more species in their enriched fallows than in their wild fallows, and only two had fewer.

While the difference in the average species richness between enriched (25) and wild (22) fallows was just three species, differences in species composition among enriched fallows was greater than among wild fallows. Biodiversity levels varied considerably with the intensity and frequency of the owners' interventions. For instance, one farmer maintained more than twice the number of species (39) than another farmer in the same size fallow (0.3 ha), while in their wild fallows they maintained a similar number of species. Of all the species sampled in the enriched fallows, 42 per cent of the species were found in just one fallow and only one species was found in all fallows.

There was also a difference in the density of individual plants per area and species richness. More individuals of species were maintained in enriched fallows (average 557 in 0.26 ha) than in wild fallows (average 347 in 0.26 ha). Differences among the sampled enriched fallows were greater than among the sampled wild fallows. For instance, one farmer maintained more than three times more individuals (1120) than another in the same size (0.3 ha) fallow. Such differences were also present in fallows owned by a single family. For instance, one farmer maintained varying densities of individuals in his three sampled enriched fallows.

Plant diversity and density of individuals and species were clearly influenced by the intensity and frequency of management operations, as well as the way in which the seedlings and saplings of valuable species were managed in the field stage. Results of floristic and structural analyses were consistent with field observations. Vegetation of wild fallows that were not managed was clearly dominated by individuals of imbauba (*Cecropia membraneaceae*) and other early-colonizing pioneer species. The majority of enriched fallows contained agroforestry species including bananas and several species and varieties of citrus. We focused a considerable amount of attention on enriched fallows because they are under intensive management and have considerable economic and ecological importance.

Field observations showed that farmers maintained individuals of economic species at different stages of growth in the same fallows and fields. There were adult pau mulato trees near other individuals of the same species in the sapling and seedling stages. These uneven-aged stands created a multiplicity of habitats for other species in the landscape.

Of the eight sampled enriched fallows only two had an estimated species richness index (see Menhinink, Chapter 8: D_{mn} = 0.93 and 0.89) of less than one, while only one fallow showed an estimated Shannon's Index (H' = 0.84) of less than one. There was a strong correlation (R^2 = 0.94) between the distribution of individuals and species in the enriched fallows.

The rates of abundance and dominance of species in fallows reflected the smallholders' intensive use and management of fallows. Species richness in the enriched fallows showed a strong correlation (R^2 = 0.88) between number of species and number of individuals. The number of trees per area increased when the number of species increased in the fallows.

Although thinning and removal of vines were the main management operations applied to fallows, we found smallholders actively adapting or developing new management techniques. This transformation and innovation was facilitated by the increased value of forest and fallow products in the markets. At both sites farmers were making small openings (*clareras*) in their fallows for planting semi-annual species such as bananas, and for transplanting seedlings of desirable species. They collected seeds of several species, such as tropical cedar and açaí palm, to broadcast in their fallows. The frequency and intensity of termite nest removal and other operations to control pests were also increasing. Discussions with farmers revealed that decisions to convert fallows into fields or forests were based on several factors, including how well production of agroforestry products, such as bananas, was faring, and on whether forest species, such as açaí palm, were dominating the stand.

Forests

In both sites we found that forest areas in farmer landholdings were the result of successive management operations that began in the field stage and continued into the fallow and forest stages. Inventories conducted in a sample of 10 ha (5 ha in Mazagão and 5 ha in Ipixuna) showed a great diversity of species.

At both sites forests showed high values of species richness and evenness. The average number of species found in Mazagão forests (51) was higher than the average found at Ipixuna (36). In contrast, the sampled forests of Ipixuna had more trees (average 1117) than those sampled in Mazagão (average 1041). These results reflect different histories of management and resource extraction. In Mazagão, people have been more dedicated to forest activities and have tended to continually enrich their forest with economically desirable species of timber, medicinal plants and fruits. Farmers in Ipixuna have practised more agroforestry and the management of forests is mainly for fruits and medicinal products rather than for timber extraction. Based on the estimated diversity indices, forests in Mazagão had higher values (average H' = 2.59) than forests in Ipixuna (average H' = 1.77). These results were very similar to the reported estimated Shannon's Index values for forest areas in other regions of the estuarine *várzea* floodplain.

While forests in Mazagão were richer in species than were those in Ipixuna, the two most commercially valued species (açaí palm and pau mulato) were among the most dominant and abundant species in both sites. This abundance indicates that people were encouraging the establishment and growth of these and other valuable species. Similarly, the presence of a high number of timber, fruit and medicinal species suggests intensity of management at both sites. Data show that smallholders also maintain low numbers of individuals of several non-commercial species. Among these are pioneer species such as *Cecropia palmata* and *Croton* species that play an important role in attracting game animals.

As in the case of managing fallows, research in forests showed smallholders to be adapting and developing innovative management technologies in response to changes in markets. The abundance and dominance of economically important species were maintained through management that promoted the regeneration of species under different light and environmental conditions. For instance, the majority of farmers conducted pre-harvest operations to avoid excessive damage to the forests, thus optimizing production. Some innovative farmers broadcast seeds or planted seedlings of valuable species before cutting timber. Most seedlings were collected from other parts of the forests, although the seedlings of andiroba (*Carapa guianensis*) were mainly produced in house gardens.

House gardens

Within their house gardens, smallholders of Mazagão and Ipixuna maintained orchards, nurseries, medicinal plants, vegetables, ornamentals, spices, grasses and vines, as well as areas for raising domestic animals. The sampled house gardens included most of these management categories and all vegetation in them was inventoried. There was little variation between sites in numbers of species and individuals maintained in house gardens. Variation became apparent when individual gardens were compared. Based on these observations, the biodiversity analyses of the sampled house gardens from both sites were pooled. House gardens were generally characterized by high biodiversity, and there was no correlation ($R^2 = -0.02$, $n = 59$) between the number of species and individuals found in each sampled house garden. The average number of species found in the 16 sampled house gardens was 17; the maximum was 26 and the minimum was 11. Similarly, results show that the number of individuals found in house gardens varied from 136 to 815. Most of the species found in the house gardens with a higher density of plants were herbs, grasses or vines planted for medicinal, ornamental, spice and food uses. The majority of species inventoried in lower density gardens were palms and trees.

While the levels of biodiversity in fields, fallows and forests were strongly dependent on the intensity and frequency of production and management technologies, the number of species in house gardens depended more on their uses. House gardens that comprised orchards, nurseries and gardens had a greater number of species than those with only one Field Type.

Conclusions

Economic activities in Amazonian environments are generally assumed to result in severe declines in biodiversity. Our research in the Amazon estuary, however, shows that human use of floodplain forests need not be a cause of biodiversity loss. Moreover, it demonstrates that economic changes and more intensive market involvement need not lead to environmental degradation.

This case study provides an example of a social group with a tradition of changing and adapting to the dynamic social and natural processes in the region. Change is a constant in all environments, and the Amazon is surely among the most dynamic. The rapid and profound effects of fluvio-dynamic processes that characterize change in the Amazon floodplain may be unusual, but the ability of local populations to respond appropriately and effectively to a variety of changes by manipulating of their environments in new and ingenious ways is not unique to Amazonia. The vegetation of the floodplains is a product of long, complex, and shifting interactions of physical, biological, economic, and social events and processes. While our research attempts to trace trends, elucidate interactions, and dispel some prevalent myths, the precise contribution of any one of the processes that built the rich biological diversity of Amazonia is difficult to detail. To underrate the contribution of any of these factors is, however, to misunderstand how Amazonian biodiversity is created, maintained, changed and destroyed.

16 From forests to fields: incorporating smallholder knowledge in the camu-camu programme in Peru[1]

MIGUEL PINEDO-VASQUEZ, AND
MARIO PINEDO-PANDURO

Background

EXPERTS HAVE RECOMMENDED the cultivation of Amazonian forest products with economic value to local farmers. This would effectively replace unsustainable patterns of resource extraction and biodiversity loss and lead to more sustainable resource management (Homma, 1992; Moran, 1993; Arnold, 1995). The steps that need to be followed and ways to motivate the *ribereños*, the smallholders of the Amazon floodplain of Peru, to plant and manage forest species in their fields, fallows, house gardens and forests have been outlined by Dufour (1990), Lampietti and Dixon (1995) and Peters (1996). Many other authors have discussed the economic and ecological advantages of cultivating rather than extracting forest products (Dufour, 1990; Afsah, 1992; Clay and Clement, 1993; Arnold, 1995; Homma, 1996).

Development and conservation agencies have acted upon this advice and are implementing programmes in Amazonia that promote the production of products that were formerly extracted from the forest. Most of these programmes are implemented by public agencies and NGOs that design and promote technical packages developed by urban-based experts on research stations. The majority of the programmes include several requirements, such as size of planting and soil type, that farmers need to comply with in order to become beneficiaries of the programme. One such activity in Peruvian Amazonia is directed toward the cultivation of a shrub species, camu-camu (*Myrciaria dubia*), for the production of the vitamin-rich camu-camu fruit. The fruit of *M. dubia* contains about 2000 – 3000 mg of ascorbic acid per 100 g of pulp (Peters and Hammond, 1990), making it the fruit with the highest known concentration of vitamin C – over 30 times that of an orange (FAO, 1986).

An increase in demand for camu-camu fruit on the international market has been predicted because of its high vitamin C content, with suggestions that production could become a viable alternative to coca (*Erythroxylum coca*) production in several regions of the Peruvian Amazon (El Comercio, 1996; Expreso, 1996). Although there are few reliable data on the amount of camu-camu fruit that is being extracted currently and sold in markets, most experts believe that the demand cannot be supplied by continuing to extract fruits from natural stands. *Ribereños* are being encouraged to plant *M. dubia* in their fields, fallows, house gardens and managed forests.

Overharvesting is believed to have a negative impact on natural regeneration of the species as well as on fish populations (Goulding et al., 1996). Camu-camu fruit are an important source of food for some of the most valuable Amazonian fish species such as gamitana (*Colossoma macroponum*). The adaptability of *M. dubia* and its tolerance of floods and other environmental constraints of the floodplain are other important reasons why *ribereños* are being encouraged to plant the shrub (FAO, 1986).

Research and extension activities promoting the planting of *M. dubia* began more than 20 years ago, and until recently have had very little success. Since 1995, fruit production has been part of a reforestation programme funded by the Peruvian government and implemented by private enterprise. Despite the considerable time, expense, and work that have been devoted to the promotion of camu-camu, *ribereños* have hardly responded to the government programmes or incentives. One programme that has incorporated *ribereño* knowledge into the technology package is promising more favourable results. In this chapter we discuss some reasons why many of the government-initiated programmes have been unsuccessful, and how using local knowledge and technologies, and the help of village experts, has led a to a better planned project, and has helped to explain the reluctance of smallholders to plant the fruit. This case is from a study of villages located in a region locally known as sector Muyuy. This sector is part of the white-water or *várzea* floodplain located near Iquitos, the largest urban centre in the Peruvian Amazon (see Figure 16.1).

What is known about ribereño management systems

Ribereños have developed complex agriculture and agroforestry production techniques suited to different *várzea* environments (Padoch and de Jong, 1987). Most have traditionally managed and planted a great diversity of forest species with a wide range of uses (Padoch and Pinedo-Vasquez, 1996). In the changeable and very risky *várzea* environment, *ribereños* have long experimented with different management schemes and species, selecting the best genotypes and the most suitable environments for their production. Studies have described a wide array of production systems, but none report farmers planting monocultures of trees or woody shrubs on levees (Padoch and Pinedo-Vasquez, 1996). These programmes have taken into account some ecological data about camu-camu and the kinds of stands it naturally forms, but have included only limited information about farmers' practices and needs. Such information is critical for planning a role for *ribereños* in the process of cultivating camu-camu or any other forest products.

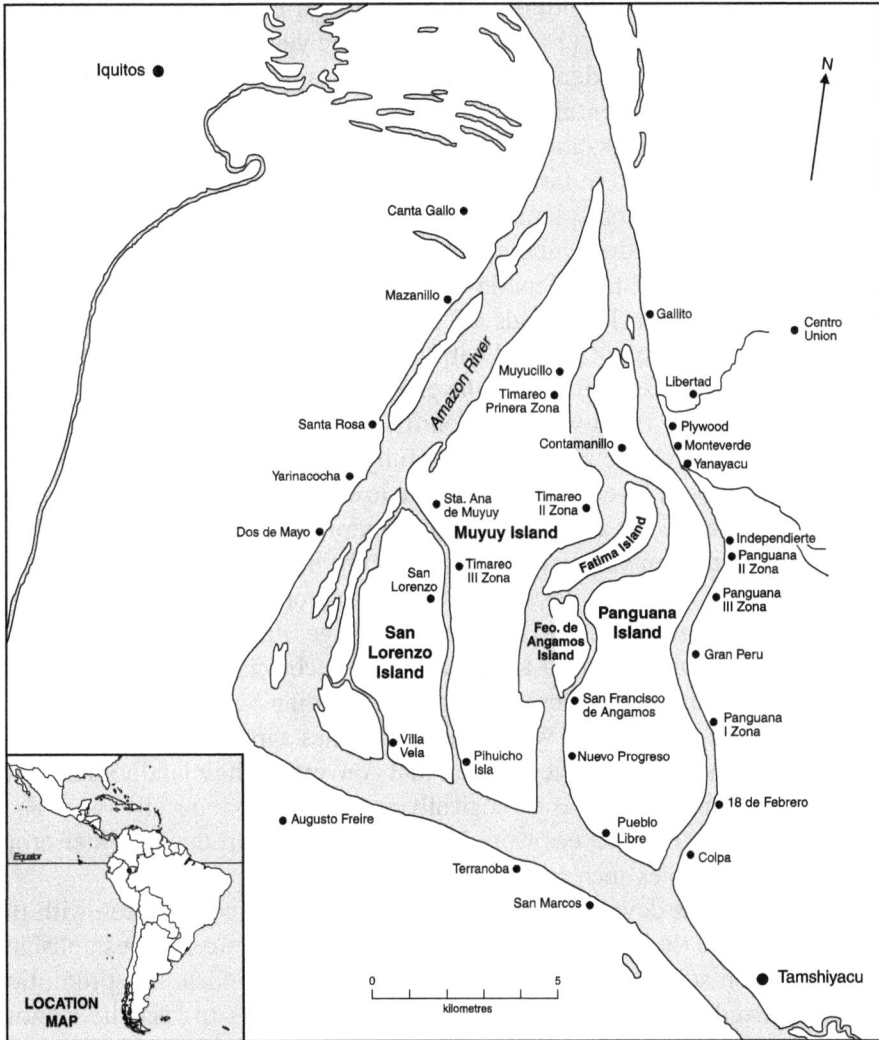

Figure 16.1   *Region near Iquitos, Peru, where the planting of camu-camu fruit is being promoted*

Motivating *ribereños* to plant *M. dubia*

The process of promoting *M. dubia* cultivation among smallholders near Iquitos has involved several researchers, urban-based technicians, and *ribereño* households. The process began in 1980 when a total of 279 seedlings of *M. dubia* were planted in an experimental plot of 0.4 ha located at the Padre Isla research station (Pinedo-Panduro, 1989). Seedlings were inter-planted with other fruit species adapted to *várzea* conditions. *M. dubia* seedlings showed the highest survival rate (100%) and were the first ones to produce fruits (Pinedo-Panduro, 1996). Although camu-camu yields in the plots were only 9 per cent (1 t/ha) of the average yield of 11 t/ha produced in natural stands, the study demonstrated that it was possible to cultivate camu-camu. Further experimental studies involved varying the distance between planted seedlings. Yields, however, failed to improve significantly (1.1 t/ha). Experiments conducted in farmers' house gardens yielded similar results. *M. dubia* seedlings were transplanted from natural stands to the lowest sections of house gardens. Fruit production obtained by these families after four years ranged from 0.7 to 1.2 t/ha (Pinedo-Panduro, 1996).

Despite the rather low yields of fruit produced in house gardens, six *ribereño* families who cooperated with the camu-camu programme reported that they were not dissatisfied with the experiments. Among the most valuable results obtained by them was the identification of the specific *várzea* areas where *M. dubia* could be planted successfully. Farmers found that it grew better on backslopes of levees or *bajeales* (the lowest section of the *várzea* lands) that remain flooded for an average of seven months a year. In addition, all six farmers found that on *bajeales M. dubia* could be interplanted with rice, maize, beans, watermelon, vegetables and other annual crops. They developed a technology significantly different from the package that had originally been promoted by the urban-based technicians. The technicians had advised farmers to plant *M. dubia* on levees (*restingas*), rather than *bajeales*, in agroforestry systems or in monocultures. While the six families agreed that fruit could be produced in monocultures, they argued that converting their landholdings into plantations of this shrub was ecologically as well as economically risky. They also decided that their levees were better devoted to producing other crops including food staples such as plantain and cassava.

Planting models developed by the six households in cooperation with the technicians were then promoted to other potential producers by establishing demonstration sites on the landholdings of the six families. The promotion process started by organizing groups from communities to visit the demonstration camu-camu fields. After the visits, families that expressed an interest in planting *M. dubia* were provided with seedlings and their landholdings were visited. The 'demonstrator households' were particularly good at giving relevant advice on production and the economic risks associated with the new endeavour. They helped their neighbours make reasonable judgements on whether they should 'self-select' themselves as good candidates for the process of experimentation and problem-solving that the camu-camu programme still required.

The experience of establishing demonstration sites in the *ribereños*' landholdings and organizing visits had varied results. Most farmers who visited the demonstration sites expressed interest in participating in the camu-camu programme, but they continued to vary their planting methods. For instance, households from the communities located in the complex of hamlets known as sector de Muyuy planted 2170 seedlings in small patches in a total area of 0.9 ha of *bajeal* lands. The technicians had planned for much larger, continuous plantings. The farmers, however, not only interplanted other crops including watermelon, maize and vegetables, but took advantage of a patchy environment to plant their *M. dubia* in the small areas that presented microenvironments best suited for cultivating the shrub. Some of the areas measured only a few metres square.

The village demonstration sites have already played a key role in the process of converting camu-camu fruits from an extracted to a cultivated product. The participation of households consenting to have their lands used as demonstration sites has provided insights on how to promote production among *ribereños*. An important lesson learned was that promotion of a new crop needs to be conducted primarily by *ribereños*; technicians can best play a secondary role. *Ribereños* trust other *ribereños* to identify who would be most suited to production of camu-camu fruits. Technicians can provide some useful technical and scientific advice on how to produce a product, but often this information is then made available to the wrong people.

Apart from the valuable lessons learned from the households on how to promote production, new and useful technical information was also obtained from these families. For instance, the farmer experimenters found that *M. dubia* seedlings in *bajeales* were vulnerable to desiccation and weed invasion. Households identified weeding as the most difficult and costly maintenance operation. However, after three years of experimentation, weed invasion was more successfully controlled by increased interplanting of maize, beans, watermelon and vegetables. Intercropping helped reduce the production costs and compensate for the relatively low yields of fruits obtained. Yields of fruits harvested by these farmers remained within the range (an average of 1.1 t/ha) of the fruit yields obtained in the experimental sites at the research station of Padre Isla.

The use of the *ribereño* fields as demonstration sites has helped expand the cultivation of *M. dubia* to other communities. This has been facilitated by the availability of financial incentives provided by the Peruvian government since 1995. The project team began working with a private institution, the Compania Amazónica de Producción Forestal (CAMPFOR) based in Iquitos. Several private institutions, including CAMPFOR, obtained grants from the government in 1995 to promote the planting of *M. dubia* as part of the reforestation programme. While most private enterprises used the grants to put in their own plantations, members of the CAMPFOR team used the grant to continue working with *ribereño* families in developing *M. dubia* as a cultivated crop. This was seen as a direct way helping smallholders increase their income. Although cooperating farmers had yet to harvest and market fruits, the project was counted a partial success because of the information that has been obtained through the partnership with farmers. Using smallholders as advisers on technical matters of production and on the best ways to promote the project in the area has proved to be both rewarding and efficient.

Discussion

Programmes promoting the planting of extracted species as a way of conserving them while enhancing rural incomes have long existed in Amazonia. Planting rubber (*Hevea brasilensis*), cocoa (*Theobroma cacao*) and other species that produce valuable commodities have been intensively promoted by private and public agencies since the end of the nineteenth century (Almeida, 1996). Urging *ribereños* to plant *M. dubia* for the production of camu-camu fruit is one of several programmes currently being implemented by NGOs and state agencies in rural communities. Most *ribereño* families living in the Iquitos region have participated or are participating in these programmes.

While projects that promote the cultivation of forest species offer some technical and economic incentives appreciated by *ribereños*, their proliferation in numbers, their short time-frames, and competition among them is creating an environment that limits rather than facilitates the participation of farmers. Most NGOs and state institutions engage only in short-term programmes (an average of two years) to promote activities whose success must be measured in the long term. Economic incentives are used to attract as many farmers as possible. Because of the short project lifetimes and the economic advantages offered, most *ribereño* families perceive these programmes as short-term financial opportunities and not as viable livelihood alternatives or ways to increase their future household income. Lack of continued technical and financial support from NGOs or state agencies was one of the main reasons given by farmers as to why some of them quickly abandoned their plantations of *M. dubia*, even some of those located in *bajeales*.

While lack of continued effective support has negative impacts on adoption, this was not the main reason why many farmers who were first offered *M. dubia* seedlings stopped participating in the projects. The package was not well suited to their management needs and only those few experimenters who were motivated to continue the process of determining acceptable planting conditions managed to persist and benefit. The ecological, economic and social factors that involve cultivating previously uncultivated species require a certain level of expertise and interest that not all *ribereños* have, and that most of the technicians could not offer. Identification of local experts who could aid technicians in determining the optimal production conditions was an important step, which was not considered by most NGOs and state agencies. The village experts eventually 'self-selected' themselves and continued the necessary experimentation.

Apart from the need to include *ribereño* experts in cultivation trials, further promotion of production techniques and methods also required the use of demonstration sites and 'demonstrators' whose information was valued and trusted by other farmers. While most NGOs and state agencies use research stations as demonstration sites, we found that the most efficient way to promote the planting of *M. dubia* was by using the farms of the self-selected households as demonstration plots. By observing and learning from these experts, other *ribereño* farmers could evaluate whether they too could overcome the technical and financial problems that would arise when NGOs or state agencies ceased to be involved. Self-selected farmers also provided valuable technical advice to their fellow farmers and to the technicians. By working with *ribereño* experts we learned that for most farmers the important questions went beyond how or where to plant *M. dubia*, to the kind of management needed to reduce damage caused by floods, strong river currents and weed invasion.

Apart from their technical input, the expert households played an important role in the identification and selection of other prospective growers of camu-camu fruit. The main concern in deciding whether to plant was the lack of information on how to overcome ecological and economic risks. Few urban-based technicians could appropriately evaluate these risks. Such critical information was provided by the *ribereño* experts and the households that had continued to experiment during the visits to the demonstration sites.

The experience of using demonstration also helped us appreciate just how greatly the urban-based technicians' plans for camu-camu differed from those of the *ribereños*. For instance, technicians expected to establish monocultural plantings, while *ribereños* wanted to plant this species as a secondary crop interplanted with other species, maintaining both plant diversity and economic flexibility. The farmers expect camu-camu eventually to be a source of some additional income for the household, while most technicians and urban-based experts expected camu-camu fruit to become the main source of income for smallholders.

While planting *M. dubia* has increased the value of areas of *bajeales* that are too low for planting crops such as banana and cassava, the production of camu-camu fruit cannot replace food staples or other crops as the main source of household income. In addition, most *ribereños* believe that it will be difficult for them to compete in the market with private enterprises that are establishing large industrial-scale plantations of *M. dubia* in several regions of Amazonia. Past experiences of promoting the production of forest products such as cocoa and guaraná (*Paullinia cupana*) support the misgivings expressed to us by *ribereños* concerning the marketing problems of smallholders. Despite the limiting factors, it is clear that *ribereños*, especially local experts and experimenters, have already played an important role in the process of converting camu-camu fruit from an extracted to a cultivated resource. They will continue to do so, as well as develop other important economic resources in Amazonia.

17 Sustainable management of an Amazonian forest for timber production: a myth or reality?[1]

MIGUEL PINEDO-VASQUEZ, AND
FERNANDO RABELO

Background

SEVERAL RESEARCHERS have exposed the limits of current methods of forest management for timber production as a means of protecting biodiversity and reducing deforestation rates in Amazonia (Robinson, 1993; Rice et al., 1997). The uncontrolled or liquidation logging practised by timber companies is mentioned as one of the reasons why sustainable management of Amazonian forests is a myth. The low density of seedlings and juveniles of valuable species such as mahogany (*Swietenia macrophylla*) and tropical cedar (*Cedrela odorata*) in logged forests have been found to be the result of this method of timber extraction (Dickinson et al., 1996; Putz and Viana, 1996).

Ecologists have long argued that timber extraction produces adverse changes in water quality and other ecosystem services. Most conservationists use such ecological arguments to explain why timber cannot be sustainably produced in biogeographic regions with high biodiversity such as Amazonia (Robinson, 1993). The dependency of mahogany and other valuable timber species on forest openings for regeneration, and their poor growth performance in mature stands, are other important reasons used by conservationists to argue against timber management as a land-use system that can reduce deforestation and biodiversity loss (Rice et al., 1997). Many researchers also believe that timber extraction, regardless of where it is done, what its scale, or by whom it is practised, causes biodiversity loss, fragmentation of habitats, and other ecological and economic problems. Selective logging erodes seed banks and greatly reduces the capacity of valuable timber species to regenerate (O'Connell, 1996; Putz and Viana, 1996). Timber extraction is a major cause of forest fragmentation and does have negative effects on the structure, density and distribution of the vegetation, and adverse impacts on habitat diversity for wildlife (Roberts and Gilliam, 1995).

There are also economic and political factors that impede the sustainable management of forests. The market is very selective and depends on mahogany and a few other slow-growing hardwood species (Plumptre, 1996). Investing time and money in the management of these species requires a long-term investment in a very unstable and risky political environment (Rice et al., 1997). The majority of governments do not have the necessary land and resource tenure policies to promote sustainable management (Barbier, 1995). The lack of tax and other economic incentives for loggers and timber enterprises to conserve is another important reason for saying that sustainable timber production in Amazonia is a myth (Uhl et al., 1997).

A close look at how smallholder (rather than industrial-scale) timber managers confront these problems and limitations offers a very different picture from the one presented by the majority of experts. We examined timber activities of smallholders in the estuarine *várzea* floodplain located in the Brazilian state of Amapá. In the landholdings and managed forests of rural Amazonians, known as *caboclos*, we found evidence of sustainable and profitable production of timber that does not substantially reduce biodiversity. In this chapter we discuss the current and past economic and political conditions that favour or limit timber production. Management practices, the reasons why smallholders are engaged in timber activities and the economic returns from timber are analysed. We also assess the ecological sustainability of the *caboclo* system of timber production by quantifying the impact of forest management techniques and extraction on species, habitat diversity, natural regeneration patterns and availability of seed banks. We discuss how the *caboclo* system can help to protect biodiversity, and whether it can serve as a model for sustainable resource use in other parts of Amazonia.

Site and household selection

The data reported in this chapter were collected from 1991 to 1997. All information was gathered from villages located in the estuarine floodplain areas formed by the Amazon, Mazagão, Ajudante and Mutuaca Rivers (see Chapter 15, Figure 15.2). In 1991, preliminary plant surveys were conducted in the fields, house gardens, fallows and forests owned by 140 households of the total 185 families that live in the area. Members of each household were interviewed about their production and management activities, including timber management. Data collected on the inputs and outputs of timber activities began in 1992 in 35 households. This sample had declined to 12 by the end of the five-year period of the study (December 1997).

Impacts of timber management and extraction were monitored on the landholdings of the 12 sampled households. Plant inventories and evaluation of natural regeneration were conducted every two years. Changes in log prices and in the volume of timber sold by each selected household were also recorded. Trees selected and protected to serve as seed producers by each family were counted, measured and the species identified. Past timber activities in the region were recorded from the archives of the municipalities of Macapá and Santana and by interviewing selected long-term residents at the site.

Why smallholder households engage in timber activities

We found ecological, economic, social and historical information that explained why smallholders of Amapá engage in forest management for timber production. Several ecological and environmental factors make the estuarine *várzea* an excellent place for managing timber. Among these is the proximity of forest areas to river channels; most *várzea* land forms are easy to reach during high tides. Although the area contains rich soils, crop production is limited because most of the land is exposed to daily tidal freshwater floods. While the floods limit crop production, they favour the establishment and growth of forests. Floods provide the main vehicle for seed dispersal and are the source of new sediments whose high nutrient content facilitates the rapid natural regeneration of tree species. The availability of nutrients and water, and the adaptability of most timber species to flood and other local environmental conditions, make the estuarine *várzea* one of the most suitable environments for timber management in Amazonia.

Historical records show that the rich stocks of commercial volume of tropical cedar (*Cedrela odorata*), samaúma (*Ceiba pentandra*), virola (*Virola surinamensis*) and muratinga (*Maquira coriacea*) have allowed a very important timber industry to flourish in the region in the past. From the beginning of the twentieth century until 1970, seven large sawmills, with daily production of 20000 m³, and four plywood factories were operating in or near the study area. These industries closed down or moved to other regions during the 1970s and early 1980s mainly because the stocks of the four dominant commercial species were exhausted.

The departing timber enterprises left behind unemployed people with skills in sawmill operation and timber processing. Although most former timber workers moved to the cities of Santana and Macapá, some of those who remained in the area built small family-run sawmills and other timber-processing plants using mainly recycled materials from the abandoned facilities. With the stocks of the most valuable timber species exhausted, the owners of these small sawmills began to buy and process species that smallholders managed in their forests, fallows and house gardens for firewood, medicine and fruit production.[2]

The majority of smallholders were already managing tropical cedar and other valuable species. The small sawmills then created a market for several fast-growing species such as pau mulato *(Calycophyllum spruceanum)*. The demand for second growth species has made forest management for timber production economically viable and produced a change from the traditional dependency on a few slow-growing species.

Other important factors have provided the impetus for farmers to engage in timber activities. Smallholders cannot compete with the crops that are produced in the south of Brazil, and therefore began producing bananas as a source of cash income. During the past 15 years, however, the region has suffered from an epidemic of moko disease (*Pseudomonas solanacearum*), a devastating bacterial disease that has practically eliminated banana plantations and left farmers with very few sources of income. Consequently, smallholders depend greatly on the production of timber on their land. To produce timber they moved from market-oriented agriculture and subsistence forestry and agroforestry, to subsistence farming and market-orientated agroforesty and forestry.

Timber management and extraction practised by smallholders

Farmers managed timber by maintaining a patchy landscape where forests, fallows, house gardens and fields provided a diversity of habitats and environmental gradients suitable for a range of types of regeneration and growth of different slow- and fast-growing timber species. Patches ranged in size, location and successional stage. While the size and location of house gardens tended to remain stable, the size and location of fields, fallows and forests varied. The average size of house garden per household was 2.7 ha. An average of two new fields per family, covering an area of 1.4 ha, was made every year from 1992 to 1997. Most fields were made in fallow rather than in old forest. During the five years of the study, each household had an average of 3.2 ha of producing fields. Maize, watermelon, vegetables and sugar cane were cultivated in fields for two years.

While the farmers were unable grow bananas in their fields, they managed to cultivate them productively in fallows using a complex agroforestry system that minimized the damage caused by moko disease. The average area of fallows of different ages (from one to five years) per family was 6.5 ha. Fallows were managed for an average of five years, after which production of banana declined. At that stage they were either converted into fields or reverted to forest. These patterns of management have led to the establishment of multi-aged forests where both fast- and slow-growing species are found. The average area of managed forest within the properties of the sampled households was 15.6 ha. The average total area of each property was 28 ha. The maintenance of a patchy landscape and fragmented forest favours rather than limits the practice of timber management.

In order to maintain available seed banks to help regenerate timber species, farmers selected and maintained seed producer trees in all four categories of land use. Seed producer trees had a diameter at breast height (DBH) of 45 cm and showed no or very few signs of insect or disease damage. Each sampled household kept as seed producers an average of nine trees of eight species in their house gardens, seven trees of five species in their fields, 26 trees of 12 species in fallows, and 32 trees of 14 species in forests. Seed producer trees included individuals of *Cedrela odorata, Ceiba pentandra, Maquira coriacea, Virola surinamensis, Carapa guianensis* and other species that had been over-exploited by loggers in the region.

The number of seed producers and species protected in forests was greater than in fields and fallows, because seed producers of most fast-growing species were selected and protected in forests. Seed producing trees of over-exploited species were mostly kept in house gardens because in the past loggers working for the timber companies had claimed ownership of trees with merchantable volume of these species growing in smallholder forests and fallows.

In fields, timber management focused on the protection of seed producer trees and seedlings that had naturally regenerated, or were planted or trans-planted. Fallow management for timber production focused on the management of juveniles, and removal of selected vines, shrubs and pioneer trees to facilitate the maintenance of diverse gradients of light and humidity for natural regeneration. In forested areas, farmers focus on the management of adult trees for high growth and diameter increment by eliminating selected emergent trees from the stands, using girdling techniques.

Each of the 12 sampled households protected, planted and managed an average of 40 timber species on their properties. Of these, 24 were fast-grow-ing and 16 were slow-growing species. Most of the species (26) produced hardwood, and 14 produced light wood. Among the slow-growing species were the over-exploited *Cedrela odorata, Ceiba pentandra, Virola surinamen-sis* and *Maquira coriacea*. Most fast-growing species were extracted on an average rotation of eight years, while the slow-growing species were on a rotation of 30 years.

Although smallholders occasionally plant or transplant seedlings, they depend mainly on the natural regeneration of the 40 species. Data on natural regeneration collected from fields, fallows and forests of the 12 sampled households showed that the majority of timber species did not regenerate in either open or closed canopy conditions, but under the specific light and humidity conditions of fallows.

Most of the 40 timber species managed by smallholders regenerated in fal-lows (18) and most species (19) had natural regeneration equal to or less than 100 seedlings per hectare. *Cedrela odorata* and *Ceiba pentandra* had the low-est density of seedlings per area (15 per ha). Seedlings of four species had a density of more than 500 per hectare: pracaxi – *Pentaclethra macroloba* (3066 per ha), *Calycophyllum spruceanum* (2110 per ha), goiabarana – *Bellucia grossularioides* (1077 per ha) and pracuúba – *Mora paraensis* (600 per ha). While *Pentaclethra macroloba* and *Mora paraensis* regenerated in low dens-ities in fallows, *Calycophyllum spruceanum* regenerated only in fields. Because of the quality, operability and durability of the wood ($0.78g/cm^3$) and the capacity to naturally regenerate and reach commercial size, on average, in eight years, *Calycophyllum spruceanum* is considered the eucalyptus of

Amazonia. Of these four species, *Bellucia grossularioides* was the only one with low commercial value.

Smallholders maintain high stocks of commercial volume timber. Each of the 12 sampled properties contained an average of 525 trees/ha (DBH = 26 cm) with an average of 1103.88 m³/ha of standing commercial volume. Each farm was estimated to contain approximately 14700 trees with an average of 30908.64 m³ of standing commercial volume.

While commercially valuable trees of most species (36) were found in forest areas as well as in house gardens, fallows and fields, all individuals of commercial volume belonging to the four timber species over-exploited by logging companies (*Cedrela odorata, Maquira coreacea, Ceiba pentandra* and *Virola surinamensis*) were found only in house gardens and fields. An average of 24 trees/ha with an average of 53.98 m³/ha standing commercial volume of these four species was found in the house gardens and fields. Over the five-year study period, the sampled households extracted none of these trees. They were protected as seed producers or kept as a source of cash for use in a time of economic crisis.

The abundance of commercial standing volume of fast-growing species helped farmers keep trees with commercial volume of over-exploited species. Two fast-growing species alone, *Calycophyllum spruceanum* and macucu (*Licania heteromorpha*), contained the highest number of trees/ha and the largest standing commercial volume. Each property in the sample contained an average of 111 *Calycophyllum spruceanum* trees/ha with 253.31m³/ha of standing commercial volume, and 92 *Licania heteromorpha* trees/ha with 352.02 m³/ha of standing commercial volume. These two species comprised 39 per cent (5684) of trees and 55 per cent (16949.24 m³) of the estimated standing commercial volume on the average property.

An average of 2324 m³/per year (83 m³/ha) of timber was extracted per household during the five-year study. Of these an average of 37 per cent (860 m³) came from *Calycophyllum spruceanum*, 23 per cent (534 m³) from andiroba (*Carapa guianensis*), 12 per cent (279 m³) from macacaúba (*Platymiscium huberi*) and the other 28 per cent (651 m³) from 26 other species. All logs extracted by the 12 sampled households during the five years were sold to the owners of small family-run sawmills.

The price per cubic metre of logs paid by the owners of small sawmills increased by three times from January 1992 (R$ 4/m³) to January 1996 (R$ 12/m³). Although the management of timber in a patchy landscape and fragmented forests included direct and indirect costs, all sampled households calculated profits from timber extraction, discounting only direct costs. Most of the indirect cost, such as making fields, was discounted from agriculture or agroforestry production. The most common direct management costs of timber production cited by the 12 sampled households were related to the management of seed producer trees, natural regeneration, juveniles and mature trees.

The average annual net revenue made by each household from selling 2324 m³/per year of logs during the five-year study was R$ 13 981/per year. In 1997 this was approximately equivalent to US$8000. Since most management and extraction costs were covered by the labour of family members, the household income was greater than the estimated net revenue. In addition, because the owners of small sawmills pay yield tax, the state of Amapá provides tax incentives to farmers to manage timber on their properties. While a further economic analysis is required, the results from the standard cost–benefit analysis shows that timber management is a viable source of household income for the *caboclo* families living in the *várzea* of Amapá.

Discussion

The search for answers on how to produce timber in a sustainable way in tropical rainforest has produced a large body of varied and important ecological, socio-economic and political information. Although such information is extremely valuable for understanding the issues of sustainability, researchers tend to have a very limited notion of forest management. Generalizations often ignore rather than help to integrate site-specific experience that can help in finding solutions to rapid deforestation and the decline of biodiversity in places like Amazonia.

Most experts are well informed on the activities of industrial-scale loggers and timber companies, but their knowledge of smallholder timber activities is inadequate. They tend to represent smallholder forest use and management as essentially, if not exclusively, oriented toward non-timber products. While non-timber products are indeed important to many tropical dwellers, the results of our study show that timber is also a very important output from forest areas that are managed for multiple uses. By diversifying their traditional land-use systems and practising multiple-use management rather than a single use such as cattle ranching, smallholders not only reduce the ecological risks but also keep open several economic options.

The ways in which smallholders manage forests for timber production, however, differ greatly from the methods used by professional foresters and industrial timber companies (Padoch and Pinedo-Vasquez, 1996). Management implemented by timber companies is based on a single system to yield a single product while smallholders manage forests for multiple products using a variety of management systems (Pinedo-Vasquez et al., 2001). Industrial timber management focuses on species exclusion while smallholders manage timber by focusing on the exclusion of selected individual trees. By removing selected trees rather than species, biodiversity is preserved in smallholders' forests. Differences between the two systems of management are usually not considered by those who argue about whether timber can be sustainably produced in Amazonia. The lack of attention given to timber management by smallholders has led to considerable misinterpretation and overgeneralization, and in turn to the formulation of inappropriate policies.

Conclusion

Several ecological studies have shown that Amazonian forests tend to recover after timber extraction. In both the short and long term, valuable tree species can regenerate in areas where timber was extracted if seed banks are left (Boyle and Sayer, 1995; Heinrich, 1995; Roberts and Gilliam, 1995). The adverse impact of current practices in timber extraction, particularly of mahogany and tropical cedar, is not in the openings that are created, but rather in the removal of all the adult individuals without leaving any reproductive trees. Such destructive extractive methods have reduced the capacity of these and other valuable species to naturally regenerate. Although most timber is extracted using these methods, there are other extractive methods that are not destructive. For instance, the method of selective weeding used by smallholders helps to eliminate vines and reduces damage by tree fall (Pinedo-Vasquez et al., 2001).

Research conducted on small-scale timber management practised by smallholders in Amazonia and Indonesia shows that biodiversity and habitat diversity increases in areas of forests that are managed for timber production (Pinedo-Vasquez and Padoch, 1996). Smallholders use diverse ecosystem management techniques where regeneration materials are abundant, creating a diverse habitat for seed dispersal and plant growth (Pinedo-Vasquez, 1995). Several important conservation applications can be found by understanding the diversity, scale and intensity of timber management that is practised by smallholders and other groups in Amazonia and other tropical regions (Franklin, 1993; Lugo, 1995; Bodmer et al., 1997).

Sustainable management of tropical rainforest for timber production is not just a matter of knowing how. There is also a need to identify and understand the historical, ecological, economic and social factors that create conditions in which timber can be sustainably produced (Putz and Viana, 1996). While there are many ecological factors that make it difficult to practise sustainable forest management, socio-economic and cultural factors have the biggest influence in the *várzea* of Amapá. For instance, most smallholders are able to practise sustainable forest management because of the high demand for fast-growing timber species created by small-scale sawmills. Ironically, these family-run timber industries were established after large timber enterprises moved from the region, having depleted the stock of the most valuable timber species from Amapá forests.

18 Diversity of upland rice and wild vegetables in Baka, Xishuangbanna, Yunnan[1]

FU YONGNENG AND CHEN AIGUO

This chapter reports two background studies in the Xishuangbanna village of Baka, one of the four demonstration sites of the China Cluster. Baka people are mostly Jinuo, members of an ethnic minority numbering about 18000 people. In former times they practised shifting agriculture only, but many villages, including Baka, were relocated closer to rivers and roads during the period 1950–80. Most now cultivate some irrigated lowland rice. The main upland crops include upland rice, maize and beans. The Jinuo used to grow about 100 types of upland rice. They still grow about 70 types to meet the requirements of households with different economic conditions, and to most effectively use lands with varied natural conditions.

In the past the Jinuo also extensively collected roots, stems, leaves, flowers and fruits of wild plants. They still collect more than 200 species of plants for food, medicine and other uses, including more than 100 species of wild vegetables, which account for a major part of the collected wild plants. Most vegetables are from forests, except for a few that are planted. Some of the planted vegetables are common in the market, and include choucai (*Acacia intsia*), baihua (*Bauhinia variegata*), xiangchun (*Toona sinensis*), ciyuantuo (*Eryngium foetidum*), shuiqin (*Oenanthe javanica*), citianqie (*Solanum indicum*) and yuxingcao (*Houttuynia cordata*).

Little has been reported on the assessment of variety diversity at the natural village level although there are some research reports on variety diversity of the fallow system at the higher administrative village level. This study began with an inventory. In the field, accompanied by the householders, researchers recorded the local names of the different varieties of upland rice, and later obtained information about the sowing time, type, shape, and other characteristics. They moved within the village through different altitudes from lowland to highland. By discussion and observation they recorded land-use history, soil fertility, and found out about existing problems and opportunities for development.

They also noted the local names of wild vegetables, and by interview obtained information about the time and frequency of collecting, tools used and the parts of plants that were collected. A survey using a structured questionnaire was designed to find out specific information such as the quantities collected, and use for home consumption and sale. The researchers also visited the local market and recorded types, quantities and prices of wild vegetables on sale.

Diversity of upland rice

The 48 natural villages of Jinuo township grew about 70 varieties of upland rice. In Baka 20 were grown, as shown in Table 18.1. These varieties can be classified into a few large categories:

○ by sowing time: early-sown, middle-sown, and late-sown
○ by variety: non-glutinous and glutinous
○ by colour: red, white and mixed colour
○ by the temperature most suitable for the variety: cold-resistant, heat-resistant and broadly tolerant.

There are two reasons for the coexistence of three temperature groups. One is that the village altitude ranges from 600 to 1250 m above sea level. Second is that new varieties are easily introduced, as this village is now near the road.

Difference in planting frequency among varieties

There were three reasons why rice varieties were planted with different frequency. Some varieties have different requirements for growth, mainly due to differences in temperature at the different altitudes. At high altitude, varieties resistant to cold were planted. When Baka village moved down from the higher altitude to the existing site, the planting area of some traditional cold-resistant varieties decreased gradually. In addition, households with relatively more lowland wet-rice land tended not to go to the uplands for shifting agriculture. This further reduced the frequency of planting of some cold-resistant varieties.

Local people have found that it was difficult to thresh two varieties, xiahong and dahong, in the hot lowland, and consequently these varieties were planted less frequently than before. They had almost stopped planting changgu, which used to be planted at the old village site, as it was not suitable to the temperature at the new village site. Landigu produced a relatively high yield on low-quality soil in the highlands, but yielded less well in the lowlands, so it was also planted less often.

Differences in the economic situations of households also determined the frequency with which rice varieties were planted. Some types of upland rice taste good, but yield poorly; only rich households planted them. Poor households preferred the varieties of upland rice that yielded well even though they may not taste as good. This ensured that they had enough rice to eat or sell in the local market.

Differences in soil fertility was the third factor determining the choice of rice variety. At the old site, per capita fallow land of Baka village was 25 mu (1 ha is equivalent to 15 mu) according to national regulations. The fallow period was between eight and 15 years, as the village was sparsely populated,

Table 18.1 Characteristics of the upland rice varieties grown in Baka, Xishuangbanna, China

Local name	Sowing time	Glutinous	Colour of shell/grain	Temperature preference	Taste	Yield	Soil fertility	Frequency
luoiii	early	G	red/red	hot	Good taste, medium round grain	high	fertile	high
liajdaogu	early	G	white/white	cold	Good taste, soft, fragrant, long thin grain	moderate	fertile	low
diancui	early	G	white/white	cold	Hard, medium round grain	moderate	fertile	low
mowanggu	early	G	white/white	hot	Best taste, soft, fragrant large long grain	high	fertile	low
sequoluo	early	G	white/white	cold	Good taste, soft fragrant, thin long grain	low	fertile	low
landigu	middle	G	red/white	cold	Bad taste, hard, short thin grain	low	poor	low
hejieba	middle	G	white/red	broad	Bad taste, medium round grain	high	poor	moderate
hebeng	middle	G	purple with yellow/white	cold	Good taste, fragrant, very large round grain	moderate	fertile	high
xiahong	middle	G	white/red	cold	Bad taste, hard long medium grain	high	poor	low
dahong	middle	G	white/red	cold	Bad taste, coarse, large, long grain with short awn	high	poor	low
xihong	middle	G	white/red	broad	Bad taste, hard and coarse, long small grain	high	poor	moderate
bailuogu	late	Non	white/white	broad	Good taste, the largest long grain	high	fertile	many
maogu	late	G	purple/red	colder	Best taste, fragrant short round grain	moderate	poor	v. low
huagu	late	G	purple with white/red	colder	Best taste, fragrant short round grain	middle	poor	v. low
changgu	late	G	white/white	cold	Good taste, fragrant, long thin medium grain	low	fertile	lowest
manyagu	late	G	purple with yellow/white	cold	Bad taste, hard, long large grain	high	fertile	high
anene	late	Non	yellow/red	hot	Good taste, large long grain	high	fertile	high
langu	late	G	red/white	cold	Good taste, soft fragrant, thin large grain	high	fertile	low
ximongu	late	G	purple with white/white	cold	Good taste, hard, fragrant, large round grain	high	fertile	low
gulala	latest	Non	purple/purple	cold	Good taste, soft fragrant, thin long grain	lowest	fertile	low

and soil fertility could recover with a long fallow period. Soil quality was high at the old site. After the community moved down to the new site, 3000 mu of fallow land was classified as part of a Nature Reserve in 1978. Per capita fallow land was reduced to 8 mu. The fallow period had to be shortened to maintain grain production. Soil fertility declined, as the fallow period was not long enough for it to recover. Consequently, some varieties, such as ximongu, which required high soil fertility, were not planted as often as they used to be at the old site and their sowing area had decreased in recent years.

Use of wild vegetables

Due to continuing uncertainties of land tenure, local people collected whatever they could find in the mountains, and there was little attempt to conserve species and varieties. Some people cut canes or trees, and even dug out plant roots. Wild vegetables of economic value and seasonal ones, in particular, were not sustainably collected. This led to degradation of wild vegetables. For instance, the top shoots of citongcao (*Trevesia palmata*) had almost all been cut and collected.

Through a long history of collecting wild vegetables, all ethnic groups in Xishaungbanna have acquired knowledge of how to collect them regularly and seasonally according to different periods of growth. Similarly, they have developed different ways of cooking wild vegetables, such as toasting, frying, and making soup or jam. Some are eaten raw. They also have various ways of processing them, including fermentation and drying. However, little has been recorded on the assessment of species diversity of wild vegetables at village level.

Results of the inventory for vegetables are shown in Table 18.2. There were 55 species of wild vegetables used at Baka. Fresh stems and leaves of plants were mostly collected. The village was very rich in resources of seasonal bamboo shoots and mushrooms and sales in the market were substantial. Both of these two wild vegetables have potential for further development. Through the sale of wild vegetables in the market, the Jinuo have increased their involvement in business considerably.

Table 18.2 Ethnobotanical inventory of wild vegetables in Baka, Xishuangbanna, China

Jinuo name	Chinese name	Latin name	Edible part	Cooking method	Season	Habitat	Frequency
Buokuoluo	Jiahaitong	*Pittosporopsis kerri*	Fruit	Boil	Jul–Aug	Primary forest	Often
Duokuolu	Caijie	*Callipteris*	Stem, leaf	Boil, fry	Apr–May	Shady, near stream	Often
Pege	Jia		Leaf	Fry, boil	Mar–May	Shady, near stream	Often
Leduolo	Huoshaohua	*Mayodendron igneum*	Flower, leaf	Fry, toast	Apr	Shady, near stream	Sometimes
Miabosele	Hongpao		Flower, leaf	Raw	Nov–Jan	Shady, near stream	Occasional
Yeduo	Lidoujian	*Typhonium diversifolium*	Stem	Fry, steam	Nov–May	Near water	Often
Duoche	Shutiejie	*Brainea insignis*	Stem	Fry, deep fry	Sept–Jun	Shady, near stream	Sometimes
Mame	Ciwujia	*Acanthopanax trifoliatus*		Toast, steam, fry	Nov–May	Shady, near stream	Often
Duo ki a	Manji	*Dicranopteris dichotoma*		Toast, steam, fry	Sept–May	Dry upland	Often
Sagang	Diannanhujiao	*Piper spirei*	Cane	Steam	Mar–Apr	Shady, near stream	Often
Gegeili				Steam	Dec–May	Shady, near stream	Often
Palienei	Yeqingcai	*Oenanthe javanica*	Leaf	Fry, boil	Apr–May	Shady, near stream or wetland	Often
Chituoluo	Xintongzhi		Leaf	Toast, fry	Nov–Apr	Dry or wet	Often
Duoge	Jieleiyizhong	*Pteridium* sp.		Soup	Jan–Dec	Near water	Sometimes
Kemo	Beifenteng	*Cissus repens*	Fruit	Raw	Jun–Aug	Dry or wet	Often
Pachele	Zhonghuaqiuhaituang	*Begonia cathayana*	Leaf	Fry, toast	Sept–Jun	Near water	Sometimes
Paga	Fengyanlian	*Eichhornia crassipes*	Leaf	Toast, fry, boil	Apr	water	Occasional
Letugemo	Yeqingcai	*Brassica chinensis*	Whole	Soup, fry	Apr–May	Shady, near stream	Often
Seche					Aug–Sept	Shady, near stream	Often
Muzhi	Neijiangjun		Whole	Fry, boil	Jul–Oct	Secondary forest	Often
Mubo	Shaozhoujun		Whole	Fry	Jul–Oct	Secondary forest	Often
Mupoluo				Boil, soup	Jul–Oct	Secondary forest	Often
Mulele	Dahongjun		Whole	Soup	Jul–Oct	Secondary forest	Often
Mupili	Lajun		Whole	Deep fry	Jul–Oct	Secondary forest	Often
Laokulu	Muer		Whole	Fry, toast	Jul–Aug	Near stream or fallow	Often
Munie	Beisheng		Whole	Soup, fry, toast	Jul–Aug	Near stream or fallow	Often
Digeye	Likeyizhong		Whole	Toast	Nov–Apr	Near river	
Sepu	Muguarong	*Carica papaya*	Stem, leaf	Boil	Feb–Mar	Near stream or fallow	Often
Tai	Ciyu	*Lasia spinosa*	Leaf, fruit	Boil	Jan–Dec	Near stream or fallow	Sometimes
Ejieena	Nuomiye	*Serobilanthus* sp.	Stem, leaf	Boil	Jul–Aug	Near stream	Often

Palin	Houpirong	Ficus callosa	Stem, leaf	Boil	Mar–Apr	Forest margin, near water	Sometimes
Yamogeye	Cijiancai	Amaranthus spinosus	Leaf	Boil, fry	Nov–Apr	Near river, field	Often
Geginie	Shuimiqi	Cardamine sp.	Whole	Soup	Jan–Dec	Near river	Often
Geginie	Qicai	Cardamine sp.	Whole	Boil	Jan–Dec	Near river	Often
Pabu	Qianjinteng	Stephania forsteri	Stem, leaf	Boil	Apr–May	Near river	Often
Paniene	Shuiqin	Oenanthe javanica	Stem, leaf	Fry	Jul–Aug	Near river	Often
Lebulu	Zhongyelu	Thysanolaena	Core	Roast, fry	Nov	Near stream or fallow	Often
Geye	Biba	Piper longum	Stem, leaf	Roast, fry, boil	Apr–May	Near river	Often
Meibu	Huangjiangyizhong	Alpinia sp.	Stem, leaf	Fry, boil	Apr–May	Near river, shade	Sometimes
Mibiabuleduo	Yutou	Colocasia fallax	Stem, leaf	Boil	Jun–Nov	Near river or shade	Sometimes
Shuonie	Choumudan	Clerodendrum philippinum	Stem, leaf	Toast, steam	Nov–Dec	Near river, shady	Sometimes
Sekuole	Yeqie	Solanum coagulans	Fruit	Fry, deep-fry, boil	Mar–May	Near river or road	Often
Duokuoluo	Shuijiecai	Callipteris esculenta	Stem, leaf	Fry	Jun–Jul	Near river	Often
Padele	Huabancai		Stem, leaf		Mar–Dec		
Bohuo	Shuixiangcai	Parabaena sagittata	Stem, leaf	Fry, boil	Jan–Dec	Near river or road	Sometimes
Geli	Kuliangcai	Solanum nigrum var.	Stem, leaf	Fry, roast	Jun–Jul	Near river or road	Often
Cuopuliualie	Nansheteng	Celastrus paniculatus	Stem, leaf	Fry, boil	Apr–May	Near river or field	Often
Biaobu	Malan	Baphicacanthus cusia	Flower	Boil	Mar–Apr	Mountain	Often
Pakuoluo	Cipao	Thladiantha sp.	Fruit, leaf	Boil	Mar–Jun	Near stream	Often
Pabuoma	Dayuantuo	Eryngium foetidum	Whole	Fry, boil	Jan–Dec	fallow	Often
Pala	Chouyunshi	Caesalpinia mimosoides	Stem, leaf	Roast, fry, boil	Mar–Aug	fallow	Often
Kaneipabuluo	Duanbanhua	Brachystemma calycinum	Stem, leaf	Boil	Feb–May	Near stream	Often
Piuge	Rouhuihujiao	Piper semiimmersum	Leaf	Toast	Jan–Dec	Rocky	Sometimes
Akiubutu	Liangmianzhen	Zanthoxylum nitidum	Leaf	Toast	Nov–Mar	Near stream	Often
Memo	Shanyo	Dioscorea alata	Root	Toast	Apr–May	Shady and wet	Often

19 Evaluation of the cultivation of *Amomum villosum* under tropical forest in southern Yunnan[1]

GUAN YUQIN, DAO ZHILING AND CUI JINGYUN

Introduction

THE CULTIVATION of cash crops under tropical forest has become increasingly popular in tropical areas of Yunnan, China. Some crops such as tea, *Calamus* species and *Baphicacanthus cusia* have been planted under natural forest for a long time by the Jinuo and Hani ethnic minorities. Other crops are more recent and include *Amomum villosum*, *Amomum tsaoko* and *Amomum kravanh*. At the beginning of the 1970s, *A. villosum*, a Chinese or Indo-Chinese species of cardamom, usually known as Chinese cardamom, and a member of the family Zingiberaceae, was newly introduced to this area. This is an important tropical medicinal plant which is highly valued. There is now more than 55 500 mu of *A. villosum* in cultivation in Xishuangbanna, which has become one of the largest growing areas in China.[2]

In Jinuo district, in the area administered from Jinghong City, there is more than 15 270 mu of *A. villosum* under cultivation and the annual production is 81.5 tonnes, which is a quarter of China's total production. It has become the major source of cash income for the local people. However, more than 3000 mu of *A. villosum* is being cultivated in Xishuangbanna Natural Reserve, with some of this being planted very close to the core zone of the reserve. It is being cultivated on a large scale before the ecological and social effects of this kind of cultivation are well known. To evaluate the effects, we conducted a series of ecological and social surveys in Baka, a village very close to the reserve, using Rapid Rural Appraisal, Participatory Rural Appraisal, participatory observation and sample surveys. The relationship between cultivation of *A. villosum* and the development of the village was analysed. The ecological effects are evaluated in this chapter and problems of resource utilization and protection are discussed.

Baka village

Baka is a village of Jinuo township in Xishuangbanna Prefecture. There are 255 Jinuo people and two Han Chinese living in the village. The Xiaomengyang–Mengla road passes through the village and winds along the Manka river. The village is 6 km from Xishuangbanna Tropical Botanical Garden. Upland and wet-rice are grown, with most of the latter along the riverbed. The mountains to the south of the river belong to the Menglun

Nature Reserve, which aims to preserve seasonal rainforest and monsoon rainforest. At the foot of the mountains close to the river there are occasional plots planted with tea, pineapple and bananas.

The first 12 households moved to the present village site from the old village 10 km away in 1971 with the mobilization and help of Xishuangbanna Tropical Botanical Garden. There were two main reasons for the move. The population of the old village was growing rapidly and transportation was poor. With the help of the Botanical Garden, some wet-rice land was reclaimed by building irrigation channels. It was expected that increasing wet-rice cultivation would reduce destruction of the surrounding forest. With a tradition of slash-and-burn farming and inappropriate policies in force during the Cultural Revolution, extensive areas of natural forest had been destroyed in order to plant food crops and rubber trees. Only a part of the forest survives at the foot of mountains and this is reserved for *A. villosum* cultivation and water supply.

Since 1971 the population of Baka has increased, although at a declining rate in recent years. A further 30 households moved from the old village in 1972. There were 164 people in Baka in 1973, 241 in 1983, but only 257 in 1993. The area of farmland per household has also become smaller. Before 1978, each villager had access to more than 24 mu of swidden land, but the amount was reduced sharply to 8 mu in 1978 when about 3000 mu of the village land was put under natural reserve. Although the government allocated 400 mu of other types of forest land to the village in 1990, each villager still owns less upland than the average level for Jinuo township, which is 23 mu per person. There are 100 mu of wet-rice land and about 2340 mu of upland in Baka. In addition, each family owns 1 mu of land for fuelwood planting. The natural forest owned by the community has not been accurately measured, but is estimated to be about 500 mu.

Food supply and cash income both depend mainly on agricultural crops. Animal husbandry is undeveloped because of disease. The major cash crops are *A. villosum* and maize. More than 50 per cent of the villagers' cash income comes from *A. villosum*. There are other crops such as rubber, tea, pomelo, passion fruit (*Passiflora quadrangularis*) and some vegetables. Rubber is a large potential cash resource for some households. In 1993, there were 413 mu of rubber plantations but most of the trees were too young to be tapped. Pomelo and passion fruit have been introduced only in recent years and their area is limited. Vegetable cultivation for sale also started in recent years and now contributes 10 per cent of household cash income. The area of tea is only 7 mu and the villagers usually consume the total production themselves.

Crop cultivation techniques are traditional. No fertilizer or pesticides are used. The yield per mu of wet-rice is 300 kg and the area is small. More than half of the grain is obtained from the uplands. However, the upland area is limited and some swidden land has already been planted to rubber, pomelo and passion fruit. Consequently, the number of years of cultivation has increased while the fallow period has become shorter. In the past, the cultivation period of swidden land was usually 1 year, and never more than 3 years, but now it is usually 5–7 years or even longer. In the past the second cultivation period began after 13 years of fallow, but now it begins after only 5–7 years of fallow, or even less. The soil has become degraded and impoverished as a result.

Cultivation and management of *A. villosum*

The area of *A. villosum* in Baka is divided into two parts. One is in the natural reserve, the other in the community-owned forest. The official data give the total planted area as 202.5 mu, but this is only the area that is liable to tax. The actual area is about 1000 mu. Cultivation begins with the farmer choosing a flat, fertile, moist place in the forest. Then shrubs and grasses are cleared. If the tree cover is too dense, some tall trees are cut until the cover is about 70 per cent. When the rains begin in April or May, seedlings are planted with 1 m spacing between the rows. In the next year, farmers only need to hoe weeds two or three times. In the third year, fruits can be harvested. *A. villosum* is a short-lived perennial plant and it can be harvested for more than 10 years. The highest yields are in the fourth, fifth and sixth years, and then the yield declines gradually. If the environment is unsuitable or management poor, the yield per mu from old plants is very low, even less than 1 kg. Rainfall is another factor that influences the yield. *A. villosum* blooms in April and, if the rainy season begins early and the rain is sufficient, the villagers will reap a good harvest.

The cultivation of *A. villosum* began in 1974 when the government agricultural officer provided seedlings. The area planted increased year by year, at first quickly and then more slowly as suitable planting sites became more difficult to find. For example, in 1984 the area increased by 140 mu, but recently, although the price of *A. villosum* rose from 45 yuan (US$ 1.00 = 8.7 yuan) to 60 yuan per kilogram, there was little expansion. Only 3 mu of new *A. villosum* was planted at the time of fieldwork in 1993. However, the returns from *A. villosum* are so attractive that it has been planted in some unsuitable places, especially in the natural reserve.

Most of the *A. villosum* in Baka is more than 10 years old; the management is extensive and the yield low. The highest yield recorded was 85.7 kg per mu. Now the average yield is about 5 kg per mu. The same problem exists in the whole Jinuo area. There was 11 000 mu of *A. villosum* in 1986 and the harvested area was 6000 mu; production was 81 tonnes. In 1993, there was 15 000 mu of *A. villosum* and the harvested area was 12 000 mu. Production was only 81.5 tonnes. The yield in 1993 was half that of 1983. The urgent problem is how to improve the yield as the area of land and forest is limited.

The influence of *A. villosum* planting on the tropical forest

Two sample areas were chosen in Baka to study the effects of *A. villosum* invasion on the tropical forest community. Sample area A is a forest where *A. villosum* is grown, while sample area B has no *A. villosum*. Both areas are at about 650 m above sea level. Quadrats were 50 m × 50 m (0.25 ha). The native vegetation type of both sites is seasonal rainforest in gullies dominated by *Terminalia myriocarpa* and *Pometia tomentosa*.

Surveys of the two sample sites showed that both the number of individuals and the number of species of trees were reduced greatly at the different height intervals in sample A compared with B, as shown in Tables 19.1 and 19.2. In sample A the number of individuals with crowns greater than 30 m in height was less than half the number in sample B, and the number of species was four compared with seven. In the 15–30 m height interval there were seven individuals in Sample A compared with 21 in Sample B and the number of species was also lower.

Table 19.1 **Number of individual trees in different height intervals**

Site	Height intervals (m)			
	>30	15–30	5–15	1–5
A	6	7	33	> 160
B	15	21	37	> 250

Table 19.2 **Number of tree species in different height intervals**

Site	Height interval (m)			
	>30	15–30	5–15	1–5
A	4	6	12	23
B	7	8	15	42

Normal regeneration of the forest has been affected by the cultivation of *A. villosum*. Tables 19.3 and 19.4 show the height distribution of species of the upper canopy found in the two sample sites. In sample B the species of the upper canopy mostly have a continuous pyramidal distribution from the highest stratum to the lowest one. In sample A the pyramidal distribution was not present. In sample A there was an absence of any individuals of these upper canopy species in height interval 5–15 m. The frequency of individuals in the understorey was also reduced. In the case of *Pometia tomentosa*, in sample A the numbers of individuals were low or absent in the four height intervals, whereas in sample B they were higher, and 20 individuals were present in the lowest height interval. There were many young trees and seedlings in sample B, but they were scarce in sample A. *Pometia tomentosa* is a threatened species and is a characteristic species of seasonal rainforest. The potential lack of regeneration in forest where *A. villosum* is grown is a cause for concern.

Table 19.3   Distribution of tree species with a mature height greater than 15 m in different height intervals in sample A

Species	Height intervals (m)				Small quadrats
	> 30	*15–30*	*5–15*	*1–5*	*2 × 2 m*
Terminalia myriocarpa	1				
Pometia tomentosa	3			1	
Litsea liyuyingi	1				
Bischofia javanica	1				1
Polyalthia cheliensis		1			
Cinnamomum chartophyllum		2		1	4
Adenanthera yunnanensis		1		1	5
Allophyllus cobbe		1			1
Ficus oligodon		1		2	
Total	6	6		5	11

Table 19.4   Distribution of tree species with a mature height greater than 15 m in different height intervals in sample B

Species	Height intervals (m)				Small quadrats
	> 30	*15–30*	*5–15*	*1–5*	*2 × 2 m*
Ailanthus fordii	1		2	6	1
Pometia tomentosa	3	5	1	20	17
Dalbergia obtusifolia	6	12	12	15	11
Gironniera subaequalis	3			1	
Antiaris toxicaria	1			2	
Mitrephora thorelii	1			18	2
Albizia crassiramea		1			
Myristica yunnanensis		1	3	2	
Macaranga denticulata		1	3	67	34
Total	15	20	21	131	65

Species and individuals of other rare and endangered plants have also been reduced since *A. villosum* has been cultivated in the forest. Table 19.5 shows that there were ten species of rare and endangered plants in the sample from sample B while there were only seven species in sample A. These seven species were less frequently represented in the height interval counts. In the shrub height interval, from 1 to 5 m, there were only three species with one individual of each in A compared with seven species and 31 individuals in sample B.

We also found that the community structure had changed greatly in sample A. Not only was the undergrowth cleared, but the trees had also been partly cleared for the cultivation of *A. villosum*. With the forest cover reduced, some light-demanding plants had invaded the shrub and herbaceous strata. In sample A the shrub stratum was dominated by plants with large leaves, such as *Musa acuminata, Colocasia gigantea* and *Alocasia macror-rhiza*. The numbers of individuals of these three species were 45, 45, and 20, respectively. They constituted 67.4 per cent of the total number of individual plants in the shrub stratum. In the herbaceous stratum of sample A, the proportion of young trees was reduced, while grass increased. A large number of *A. villosum*, and also *Eupatorium coelestinum, Chromalaena odorata* and *Elatostema macintyrei* had appeared. In the same stratum of sample B, most plants were the young individuals of taller growing trees. For example, in two small quadrats 91 per cent of individuals and 87.5 per cent of species were young trees, but the respective percentages in sample A were 50 per cent and 33.3 per cent. More seriously, the young trees in sample A would be cleared in the next hoeing season.

Table 19.5   The rare and endangered plants at each height interval at the two sample sites

Species	Number of individuals in height intervals (m)									
	Sample A					Sample B				
	> 30	15–30	5–15	1–5	Small quadrat	> 30	15–30	5–15	1–5	Small quadrat
Horsfieldia tetratepala		1						1		
Myristica yunnanensis							1	3		
Homalium laoticum			1						1	
Pterospermum menglunense								1	4	
Tetrameles nudiflora			5							
Antiaris toxicaria						1			2	1
Laportea urentissima			1				3			
Pometia tomentosa	3		1			3	5	1	20	17
Terminalia myriocarpa	1			1						
Toona ciliata			2						1	
Mangifera sylvatica									2	
Magnolia henryl									1	
Total	4	0	8	3	1	4	9	6	31	18

There was also great change in the mid-strata plants. In sample B, there were many large woody vines, including *Acacia pennata, Combretum yunnanense* and *Tetrastigma planicaule*. In sample A, however, the common vines were small herbaceous species such as *Thunbergia grandiflora*, and seedlings of the woody vines could be seen only occasionally.

Discussion

Cultivation of *A. villosum* under natural forest has brought significant changes in the structure, species composition and regeneration capacity of the forest community. The survival and development of some rare and endangered plants are threatened. Planting *A. villosum* under forest, at least as it is currently practised, is incompatible with retaining the forest over the long term.

The social and ecological surveys reveal that the cultivation of *A. villosum* in natural tropical forest raises complex issues about the relationship between resource conservation and utilization. Our conclusion is that cultivation should cease in the natural reserve, because of the extent of the changes in the structure, composition, and characteristics of the tropical forest and its ability to regenerate. The main purpose of Xishuangbanna Natural Reserve is to protect the seasonal and monsoon rainforests, but now only 6 per cent of the area in the reserve still has these two forest vegetation types. The rich species resource of the tropical forest should be conserved, and the natural reserve should provide sanctuary for the many rare and endangered species. Cultivation in the buffer zone of the reserve should not be permitted either, because most of the forest in the zone has already been degraded to different degrees and is now regenerating. Human interference should be minimized, so that the forest can regenerate rapidly and well.

In areas surrounding the reserve, like Baka village, *A. villosum* could continue to be planted, but cautiously. Cultivation under community-owned forest has to continue in the short term, because of the present economic and social situation. More conservationist and productive techniques should be studied and adopted. To ensure regeneration, farmers should keep enough young individuals of the taller-growing trees when they plant *A. villosum*, or hoe the weeds.

A comprehensive plan for the cultivation of *A. villosum* needs to be made at village level. Only appropriate planting sites should be chosen. It should be possible to sustain market production, without further increasing the area, if the yield per mu can be raised, especially by the renovation of old *A. villosum* areas. Improved techniques of food production would increase self-sufficiency, and the development of alternative cash crops would reduce dependence on *A. villosum* and the pressure on the remaining forest.

20 Diversity at household level in wet-rice fields at Daka, Xishuangbanna, Yunnan[1]

FU YONGNENG, GUO HUIJUN, CHEN AIGUO
AND CUI JINGYUN

Introduction

THE IMPORTANCE of agricultural diversity (agrodiversity) for biodiversity conservation and farmers' livelihoods has been widely advocated (Padoch and de Jong, 1991; Brookfield and Padoch, 1994). Much work has been done at the landscape level concerning the effect on biodiversity of different land-management patterns in Xishuangbanna. Further research at household level is important because peasant households are the basis of sustainable rural development (Guo et al., 2000b). Wet-rice fields were selected for detailed study of plant composition, diversity and utility because they are particularly important agricultural ecosystems. Moreover, they sometimes contain considerable biodiversity, contrary to a general impression that they have very little. Using the HH-ABA method described by Guo et al. (Chapter 7), the authors studied the wet-rice fields of Daka, Xishuangbanna, one of the demonstration sites of the PLEC project, as a case study.

The study area

Daka is a Hani village that belongs to the Daka district of Menglun town, Mengla county, in Xishuangbanna Prefecture. There are 53 families and 304 people in the village. It is about 8 km away from Menglun town and 10 km from the Menglun national nature reserve. The average air temperature is 21.5°C, the annual rainfall is 1563 mm, and the annual mean relative humidity is 83 per cent. The soil is lateritic red earth. The natural vegetation type is tropical seasonal rainforest.

Field methods

The methodology was based on the PLEC guidelines (Zarin et al., 1999; see Chapter 6) and HH-ABA (Guo et al., 2000b; see Chapter 7). The sample size was 12 households. They were chosen randomly from the residence booklet of 53 households, and with some suggestions from the local residents. The sampling percentage was 23 per cent. One sampling plot was selected from each of the 12 households. Selection of the sampling plots within the household holding was biased, and the location was determined by the distribution and shape of the land owned by the 12 sampled households. Four 1 × 1 m quadrats were selected from each of the 12 sampling plots. As a result, there are 12 sampling plots and 48 quadrats. The habitat characteristics of all 12 sampling plots were recorded. At the same time the 12 households were interviewed to collect information on rice varieties, yields, management practices and utilized plants.

Data analysis

The species diversity of different household farms and of different sampled rice fields was calculated using the methods of Margalef's species diversity index ($D_{mg} = (S - 1)/\ln N$, where S refers to number of species and N refers to number of individuals). An agro-species diversity index was modified from Margalef for use in this case study. For calculation of the modified agro-species diversity index, S refers to number of utilized species, and N refers to number of individuals of utilized species in the above formula. The number and percentage of utilized species were calculated (Guo and Wu, 1998). We also conducted similarity analyses between different households using the Whittaker index (βw) and Jaccard's coefficient index (S_j) (Ma, 1994; Guo and Wu, 1998). The species–area curve was used to determine the minimum number of sampling households for the wet-rice field type.

Species richness and utility

The species richness and utility of different sampling plots

The species diversity index is a simple and useful measurement of diversity. Sampled households differ significantly in terms of the Margalef's species diversity index, the modified agro-species diversity index, and the utilized species count (see Table 20.1). For example, the average of Margalef's species diversity indices of wet-rice fields among the 12 sampled households is 1.67, but the maximum value of the index is 81 per cent more than average and the minimum is 26 per cent less than the average. At the same time, the average of the agro-species diversity indices of wet-rice fields among the 12 sampled households is 0.15, while the maximum is 253 per cent more than the average, and the minimum is zero. This is because there are differences in knowledge and use of plant species between households. For example, Farmer 2, Farmer 6, and Farmer 10 all use *Eichhornia crassipes* as a wild vegetable, while Farmer 11 does not use it and only Farmer 6 knows its local name. As another example, Farmer 4 and Farmer 7 both know the local name of *Dichrocephala integrifolia*, but only Farmer 4 uses it.

Importance value indices of species in wet-rice fields

Unsurprisingly, rice has the highest importance value (Table 20.2). There are some other dominant species with water-loving or water-tolerant characteristics, such as *Rotala rotundifolia*, *Marsilea quadifolia* and *Ludwigia octovalvis*.

Table 20.1 Species diversity comparison of wet-rice fields of households in Daka, Yunnan, China

Household	Area (m²)	No. of species	No. of individuals	No. of utilized species	No. of utilized individuals	% utilized species	% utilized individuals	Margalef's species diversity index	Agro-species diversity index
1	4	8	82	1	38	12.5	46.3	1.59	0.00
2	4	9	101	2	49	22.2	48.5	1.73	0.26
3	4	8	126	1	35	12.5	27.8	1.45	0.00
4	4	15	101	2	47	13.3	47.0	3.03	0.26
5	4	6	56	1	43	16.7	76.8	1.24	0.00
6	4	7	80	2	34	28.6	42.5	1.37	0.28
7	4	9	118	1	40	11.1	33.9	1.68	0.00
8	4	7	85	1	44	14.3	51.8	1.35	0.00
9	4	6	56	1	18	16.7	32.1	1.24	0.00
10	4	7	84	2	43	28.6	51.2	1.35	0.27
11	4	9	116	2	51	22.2	44.0	1.68	0.25
12	4	12	108	3	45	25.0	41.7	2.35	0.53
Field Type	48	31	1113	5	486	16.1	43.7	4.28	0.65

Table 20.2 Importance value indices of species in wet-rice fields of households in Daka, Yunnan, China

Species	Number of individuals	Relative density	Relative frequency	Importance value
Oryza sativa	466	41.87	11.93	53.8
Rotala rotundifolia	231	20.75	7.95	28.7
Marsilea quadrifolia	98	8.81	7.95	16.76
Ludwigia octovalvis	46	4.13	5.96	10.09
Cyperus iria	9	0.81	8.95	9.76
Ageratum conyzoides	75	6.74	2.98	9.72
Ludwigia adscendens	55	4.94	3.98	8.92
Eichhornia crassipes	19	1.71	3.77	6.68
Sagittaria sagittifolia var. sinensis	13	1.17	3.98	5.15
Paspalum thunbergii	11	0.99	3.98	4.97
Enydra fluctuans	19	1.71	2.98	4.69
Bulbostylis barbata	6	0.54	3.98	4.52
Ludwigia prostrata	4	0.36	3.98	4.34
Paspalum paspaloides	10	0.9	2.98	3.88
Spilanthes paniculata	10	0.9	2.98	3.52
Hypericum japonicum	5	0.45	2.98	3.43
Kyllinga brevifolia	3	0.27	2.98	3.25
Tribulus terrestris	10	0.9	1.99	2.89
Alternanthera sessilis	2	0.18	1.99	2.17
Mazus pumilus	2	0.18	1.99	2.17
Dichrocephala integrifolia	2	0.18	1.99	2.17
Hedyotis diffusa	7	0.63	0.99	1.62
Polygonum hydropiper	4	0.36	0.99	1.35
Commelina communis	1	0.09	0.99	1.08
Oxalis corniculata	1	0.09	0.99	1.08
Sphaeranthus senegalensis	1	0.09	0.99	1.08
Dichondra repens	1	0.09	0.99	1.08
Paspalum conjugatum	1	0.09	0.99	1.08
Total	1112	100	100	200

Comparative analysis and similarity analysis

Whittaker index (β_w) and Jaccard's coefficient index (S_j)

Table 20.3 shows that there were great differences among the plant communities in the wet-rice fields of the sampling households, and the similarity indices were low (> 0.56). For example, the similarity index between Farmer 5 and Farmer 11 was 0.06; on the other hand, the difference index was 0.87.

Table 20.3   Comparison of β_w and of Sj measured by binary data on wet-rice fields of households in Daka, Yunnan, China

No. of sampled households and plots	1	2	3	4	5	6	7	8	9	10	11	12
1	–	0.65	0.75	0.48	0.71	0.29	0.53	0.6	0.57	0.47	0.53	0.6
2	0.21	–	0.41	0.58	0.73	0.38	0.78	0.38	0.47	0.38	0.56	0.52
3	0.14	0.42	–	0.65	0.57	0.6	0.76	0.47	0.71	0.6	0.76	0.7
4	0.35	0.26	0.26	–	0.52	0.55	0.42	0.82	0.52	0.55	0.58	0.06
5	0.17	0.15	0.27	0.31	–	0.69	0.6	0.85	0.67	0.54	0.87	0.67
6	0.5	0.29	0.25	0.28	0.18	–	0.75	0.57	0.54	0.29	0.5	0.68
7	0.31	0.2	0.13	0.41	0.15	0.14	–	0.88	0.6	0.63	0.56	0.71
8	0.25	0.45	0.36	0.1	0.08	0.27	0.07	–	0.69	0.71	0.75	0.47
9	0.27	0.5	0.27	0.31	0.2	0.3	0.25	0.18	–	0.23	0.73	0.67
10	0.36	0.45	0.25	0.38	0.3	0.56	0.23	0.17	0.44	–	0.63	0.47
11	0.31	0.29	0.13	0.26	0.06	0.33	0.29	0.14	0.15	0.23	–	0.71
12	0.25	0.31	0.18	0.23	0.2	0.19	0.17	0.29	0.2	0.36	0.17	–

Note: The data above the diagonal are Whittaker index (β_{ws}) to indicate the difference in species composition between two plots, and those below the diagonal are Jaccard's coefficient index (S_j) to indicate similarity in species composition between two plots.

The hierarchical agglomerate graphics of wet-rice fields also indicate that plant community similarity between different households was low (Figure 20.1). Correspondingly, there was a great difference in plant composition among different households. One reason was that some households cultivated traditional varieties. For instance, Farmer 7 cultivated a traditional rice variety called baitiangu. Some households cultivated hybrid varieties. Another reason was that there were different management practices of the same rice variety among households. For example, the planting density for the hybrid varieties was very different between Farmer 9 and Farmer 11.

Species–area curves

The minimum size of a sample must represent all components of the community sufficiently, thus indicating that the sample was of sufficient size. The minimum sample size will differ according to community structure, but the sampling percentage of HH-ABA is related to the target characteristics. To determine the minimum sampling percentage of this land-management pat-

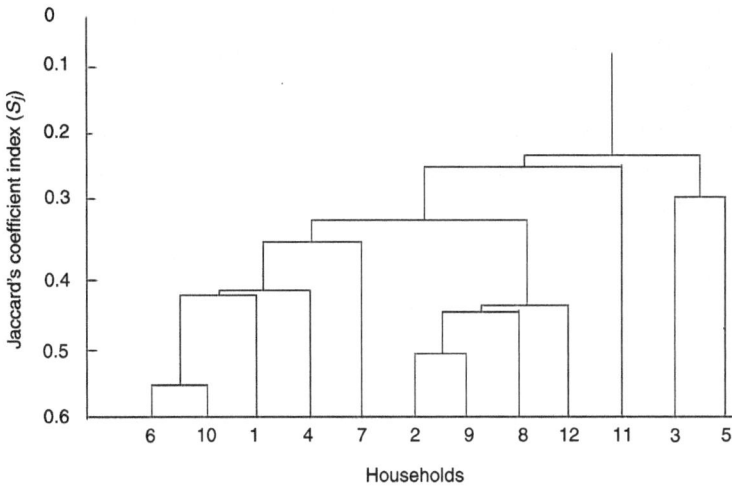

Figure 20.1   *The hierarchical agglomerate graphics of coefficients of wet-rice fields among 12 different households in Daka*

tern, the wet-rice field, we increased the number of sampling households gradually to draw species–area curves. The minimum sample percentage is reached when the species number stops increasing with increasing number of households. In this case, the point of the curve was reached when there were ten households sampled (Figure 20.2). Furthermore, the species–area curve well fitted the curve $S = b + a\ln B$. Ten households were sufficient to capture the variation in plant species diversity within wet-rice fields. The minimum sampling percentage of households for plant biodiversity inventory in wet-rice fields therefore was 19 per cent, calculated by dividing the number of sample households by the total of 53 households.

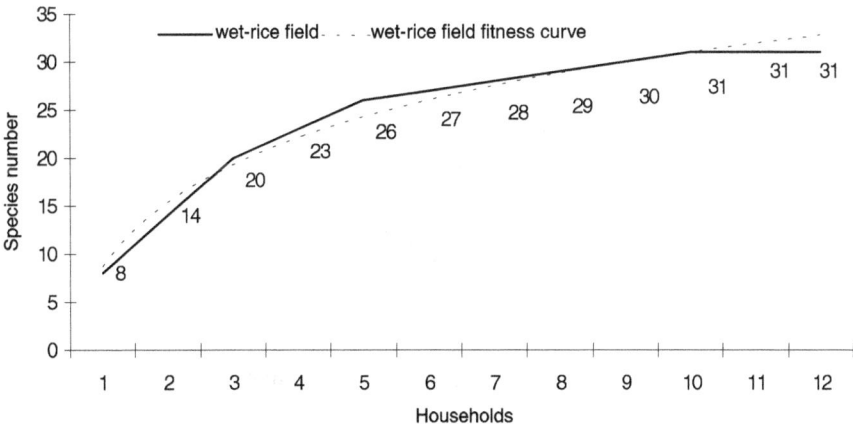

Figure 20.2   *The household–species curve for wet-rice fields in Daka*

Economic value analysis

The assessment of economic value of biodiversity has become an important topic. Comparing the diversity index above with socio-economic data for the sample households, the authors found that economic value of the biological resource in wet-rice fields varied substantially among households (Figure 20.3). The market value of non-rice utilized species in the wet-rice fields is not available, and the economic value in this case study refers to the market value of rice only. Therefore, the economic value measures the impact of germplasm selection, cultural methods, soil, and water management, as well as the effect of competing or supporting biodiversity on rice production. Despite this limitation, Figure 20.3 demonstrates that the two farmers obtaining the highest economic values sustained relatively low biodiversity in their fields, while most of those who kept or developed high biodiversity obtained low economic returns. Farmer 12, who had the second-highest Margalef Index and highest modified agro-species diversity index, had the second lowest economic value from his production. However, as the cases of middle-ranking farmers 11, 3, 4, 1 and 2 demonstrate, there is no simple inverse correlation of economic value with biodiversity. What is clear is that only a proportion of Baka farmers manage their fields so as to gain the highest economic value from their rice crop. The utilized non-rice species are clearly only of supplementary value. In so far as maximization of rice production is the general objective, there is considerable need for learning from the more successful farmers.

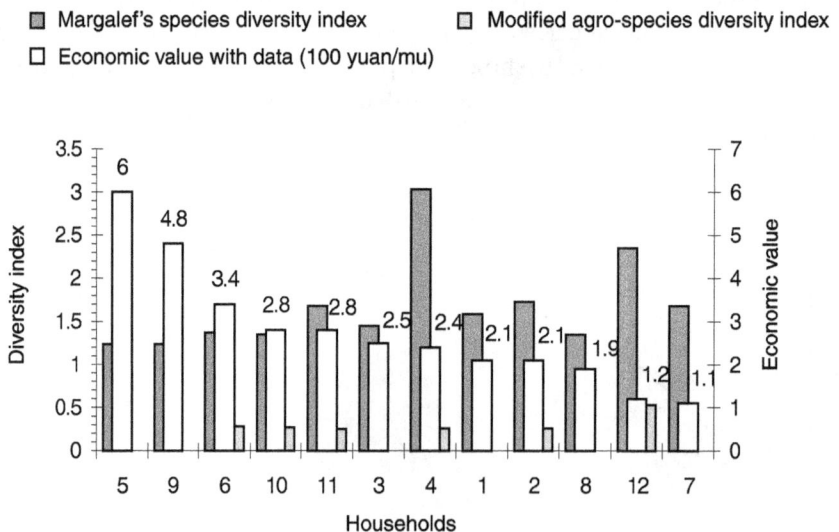

Figure 20.3   *Economic value and diversity indices of wet-rice fields of 12 households in Daka*

21 Promoting sustainable agriculture: the case of Baihualing, Yunnan

DAO ZHILING, DU XIAOHONG, GUO HUIJUN, LIANG LUOHUI AND LI YINGGUANG

PLEC-CHINA WAS one of the early Clusters working with farmers on demonstration activities before GEF funding in 1998. Activities started in Baihualing village, one of four research sites in Yunnan, China. With GEF funding the village became a formal demonstration site. This chapter describes agrodiversity in Baihualing and reviews the experiences, lessons learned and important changes in our methods during the eight years working on demonstration activities in the village. In the beginning we sought to meet farmers' needs by something akin to an extension approach, but this neglected the farmers' own roles in developing sustainable agroforestry systems. Some locally developed agroforestry is biodiversity-rich and productive. In 1999 we moved to a farmer-led approach, emphasizing the role of innovative farmers in diffusion and improvement of technologies.

Baihualing and its agrodiversity

Baihualing Administrative Village is located on the eastern slope of the south part of the Gaoligongshan Mountains in western Yunnan. The slope is on one side of the valley of the Nu Jiang River, the upper part of the Salween River. The village lies between 850 and 2000 m above sea level, and the upper part borders the State Nature Reserve of the Gaoligongshan Mountains. The reserve is well known for the richness and uniqueness of its flora and fauna (Li et al., 2000). An ancient path known as the Southern Silk Road goes through the village from China to Burma and India. Due to its special location in a biodiversity-rich environment and long history of cultivation, Baihualing is a good example of high agrodiversity.

The land area of Baihualing village is 27150 mu (Table 21.1).[1] The population is 2180, including the Han, Bai, Lisu, Yi, Hui, Zhuang, and Dai ethnic groups (Table 21.2).

Fieldwork is mainly carried out in Hanlong, which is one of ten natural villages in Baihualing Administrative Village. There are 48 households in Hanlong with a population of around 200. The Hanlong community is located in the highest part of the Baihualing village, and has an area of 1930.5 mu including a large area of forest (1500 mu) and upland fields (180 mu). House garden area is 25.5 mu. The wet-rice fields, comprising an area of 225 mu, are located 3–5 km away from the village proper, in the far valley of the Zhaotang stream. It takes a considerable amount of labour to manage these fields.

Most upland fields are distributed around the village, while the community forests are distributed above the village and bordering the Nature Reserve. There are seven major Land-use Stages and 22 Field Types in Hanlong. A general description of these is provided in Table 21.3.

Table 21.1   Land types and area of Baihualing

Land type		Area (mu)
Community forest		15 300
Farm land	Wet-rice	2115
	Upland	1200
Upland newly opened in community forest		4035
Fallow		3150
Settlement		810
Road		225
Water		315
Total		27150

Table 21.2   Ethnic groups of Baihualing and Hanlong

Ethnic group	Population	
	Baihualing	Hanlong
Han	1165	132
Bai	394	26
Lisu	367	22
Yi	161	15
Hui	81	0
Zhuang	9	0
Dai	3	1
Total	2180	200

Demonstration activities: the extension phase

Economic development remains the foremost objective of the Baihualing people because they are poor. However, the boundary of the State Nature Reserve of the Gaoligongshan limits extensive development of agriculture, which is the village's main livelihood. The only option is agricultural intensification, as the village is remote from any cities. For many years, sugarcane has been an important cash crop in the village and, with the support of a nearby sugar factory, cane fields have expanded into the uplands. The expansion was detrimental to both biodiversity and soil fertility. Some plots sampled in natural forests for the initial biodiversity inventory have since been completely cleared and changed into sugar or coffee plantations.

Table 21.3   Land-use Stages and Field Types of Hanlong village

Land-use Stage	Field Type	Management
Natural forest	FT1 Timber forest	Upper part of community forests close to Nature Reserve, usually for timber
	FT2 Fuelwood forest	Middle part of community forests for fuelwood
	FT3 Scenery forest	Close to village and near the road or path
Cultivated forest	FT1 Chinese fir forest	Introduced timber trees, usually planted on steep sloping upland fields or around them, young trees intercropped with maize before shade formed
	FT2 *Phoebe puwenensis* forest	Native timber species, naturally growing around 1800 masl, planted by local farmers more than 30 years ago, seedlings grown by farmers and planted on slightly steep fields
	FT3 *Toona ciliata* forest	Native tree, seedlings obtained from old tree sprouts, cultivated on sloping farmland or in home gardens
Agroforests	FT1 Chestnut	Usually intercropped with maize, beans, planted on upland fields
	FT2 Persimmon	As above
	FT3 Walnut	As above
	FT4 Coffee	As above
	FT5 Tea	Monocropping or as above
House gardens	FT	Trees planted on the edge, herbs and vegetables in the garden
Fallow	FT	Used for pastures, very small area
Annual cropping on upland field	FT1 Maize intercrop	Intercropped with other annual crops in summer
	FT2 Pea monocrop	In winter
	FT3 Sugarcane monocrop	Change to maize after three years
Annual cropping on wet-rice field	FT1 Wet-rice monocrop	In summer
	FT2 Sugarcane monocrop	Change to wet-rice or other annual crops after three or four years
	FT3 Potato monocrop	In winter
	FT4 Wheat monocrop	In winter
	FT5 Maize intercrop	Both in summer and winter
	FT6 Tobacco monocrop	In winter

Alternative cash crops and good practices urgently needed to be promoted in order to counteract expansion of sugarcane monoculture and to diversify cash cropping for sustainable development (Guo et al., 2000a). There is a long tradition of agroforestry practices in this area, and it was an obvious idea to promote profitable agroforestry through relevant demonstration activities.

The approaches to demonstration in Baihualing have evolved since 1995 when demonstration started with support from a MacArthur Foundation project. The initial approaches included both demonstration farmers and something rather like the standard extension programme that emphasized the role of technicians in teaching about new crops and management techniques. The combined approach met the needs of many farmers who were eager to learn about new cash crops and their management. The technicians also gave special instruction on intercropping of cash crop trees with annual crops.

Several training courses on techniques of growing fruits such as grafting, pruning, and prevention and cure of plant diseases were organized to help farmers expand areas of coffee, orange, chestnut and persimmon as alternatives to sugarcane. The expansion and diversification increased farmers' income and reduced risks arising from market fluctuations. Table 21.4 summarizes the results achieved.

Table 21.4   Area and production of main cash tree crops in 1997 and 2000 in Baihualing village

Crop	Area (mu)		Yield (kg)	
	1997	2000	1997	2000
Walnut	35	150	500	3200
Chestnut	50	130	600	4400
Coffee	82	1272	1200	18500
Longan	10	480	1	120
Persimmon	4	4	80	93
Orange	43	42	172	970

Note: Most of the cash crops are still young seedlings. The data are from village statistics. PLEC data indicate that the village statistics underestimate the area under tree crops, which now totals more than 5000 mu.

Coffee has become an important new cash crop in Baihualing due to its relatively high profitability. On average, the income from coffee is 1100 yuan (about 8.3 yuan = US$1.00) per mu, while it is only 800 yuan per mu for rice. With project support, one farmer has experimented with intercropping coffee with vegetables and fodder crops, in contrast to the monoculture of coffee by other farmers. He put 30 mu of his contracted and rented fields together for a coffee plantation and has made good money. As coffee is a new crop, most farmers need some time as well as external advice to manage it well.

With support and technical guidance, another farmer has developed an intercropping system of chestnuts, peach, maize and peas on steep uplands. The productivity of the agroforestry systems is much higher than the former cropping patterns (see Table 21.5).

Expert farmer-led demonstration

The initial approach neglected the farmer's role in developing sustainable agroforestry systems. Since 1999, household surveys and agrobiodiversity assessment have identified several innovative farmers and practices in

Hanlong and other communities in Baihualing. These farmers are very experienced in management of cash crops, and some of them have developed sustainable agroforestry systems, which they were encouraged to demonstrate. In contrast to the extension approach, the modified approach emphasized the role of innovative farmers in diffusion and improvement of technologies.

Table 21.5   Comparison of productivity between monocropping and agroforestry in Hanlong village

Farming type		Yield (kg/mu)	Value (yuan/mu)	Total value
Chestnut, maize	Nuts	45	360	
and beans	Maize	120	96	
agroforest	Beans	40	48	504
Maize and beans	Maize	150	120	
intercrop	Beans	50	60	180
Chestnut monocrop		30	240	240
Maize monocrop		150	120	120

Note: Chestnut tree density was 15 trees/mu, 5 years age; local market price.

Farmers differ in the management of biodiversity. Nine expert farmers were selected on the basis of their farming skills in 1999 and one was added in 2000. Expert farmers were categorized as generalist farmers and specialist farmers. Generalist farmers have multiple skills in management of soil and diverse cash crops. Specialist farmers have one or two special skills in management of soil or cash crops. Details about the ten farmers are provided in Table 21.6.

Table 21.6   Expert farmers and their skills

Farmer	Demonstration type	Natural village	Main skills
1	Generalist	Hanlong	Walnut, chestnut, *Amomum*, coffee and honeysuckle plantation, home garden, grafting
2	Specialist	Hanlong	Walnut and chestnut agroforestry systems management
3	Specialist	Da Yutang	Grafting
4	Generalist	Da Yutang	Upland management and timber tree plantation
5	Specialist	Mang Gang	Nursery and seedlings
6	Specialist	Mang Gang	Chinese fir plantation
7	Specialist	Gu Xinzhai	Home garden
8	Specialist	Baihualing	Dry fruit tree plantation
9	Specialist	Mang Gang	Fruit tree plantation
10	Specialist	Tao Yuan	Native timber tree plantation

Farmer 1 is one of the most expert farmers and manages his farmlands diversely and well. He was also one of the first farmers to convert his annual cropping upland to a perennial biodiversity-rich system after the individual allocation of former community land under the Household Production Contract System in 1982. A plot of 7.5 mu that his household received in 1982 was used for annual crops such as maize and beans. Because this plot was too rocky and steep to continue farming annual crops, he planted different trees such as walnut, chestnut, Chinese fir (*Cunninghamia lanceolata*), *Zanthoxylum*, flowering quince and bamboo. Some wild species regenerated naturally, such as *Phoebe puwenensis*, *Pinus armandi* and *Lindera communis*. There were more than 100 species in his agroforestry plot at the time of the biodiversity survey.

This farmer also does a great deal of experimentation in his small but diverse home garden. There were 73 species in his garden, and 71 per cent were useful species. He had grafted many pear, persimmon and new apple varieties on to local varieties of trees, and had prepared grafted seedlings of walnut, chestnut and cardamom (for medicinal use). He tells young farmers that when cultivating a tree crop, planting counts for only 30 per cent of the effort, while management absorbs 70 per cent. He often discusses grafting techniques with other farmers and teaches the younger generation both in his house and in his fields. He has a large family and is now one of the most prosperous farmers in Hanlong.

Farmer 10 is an expert in tree plantation. *Phoebe puwenensis* is a preferred native timber in Baihualing. His household received 2 ha of upland from the cooperative for maize in 1982. Since 1983, Farmer 10 has prepared seedlings and transplanted them and other timber trees in his uplands. In 1995, he rented about 30 mu of land from other villagers for expansion of his native tree business. He also started to provide seedlings of *Phoebe puwenensis* to other farmers, and began to train them in plantation techniques with support from both PLEC and the MacArthur Foundation (Guo et al., 2000b). At present, all of his uplands are covered by timber trees and other cash crops including *Amomum villosum*. He also propagates coffee seedlings.

Usually, farmers exchange their ideas informally. It is difficult to organize a large meeting of farmers. It is more convenient for them to meet in the fields accidentally and exchange their ideas. Expert farmers demonstrate their skills in the fields, not in the house, and are happy to share their knowledge with other farmers. For example, Farmer 2 helps four farmers with land around his chestnut agroforestry plot, teaching them management techniques. Farmer 5 began cultivating a traditional rice variety instead of hybrid rice two years ago. More and more farmers are following his example and cultivating this variety because of its high quality and high market value.

The Gaoligongshan Farmers' Association for Biodiversity Conservation was established in Baihualing in 1995 to coordinate biodiversity conservation and sustainable rural development. It provides a bridge between government departments (State Nature Reserve) and communities, and between projects and farmers. The Association began in Hanlong but has now expanded to include members in other natural villages. Membership is 108 and ten members are women. Several training courses in practical agricultural techniques and biodiversity conservation have been organized. New roles of the Association have been to identify expert farmers in agrobiodiversity management, and to take an important part in the organization of demonstration activities.

Conclusion

Although sugarcane has been the dominant cash crop in Baihualing village, and a large part of upland fields on the slope below the Nature Reserve was devoted to sugarcane and maize in rotation, the pattern is changing. The sugar price is unstable, and the factory demands high quality sugarcane that should be grown below 1300 m above sea level. Some risk-taking farmers and expert farmers have successfully developed profitable agroforestry systems of fruits, timber and coffee by replacing sugarcane on the uplands above 1300 m. PLEC demonstration activities, including extension of new crops and techniques, and farmer-to-farmer exchanges, have facilitated this gradual diffusion of environmentally friendly and profitable systems in Baihualing village. The leadership of Baihualing village has remarked that the project, especially the farmer training courses, has promoted the development of cash crop tree plantations, which have increased farmers' income and the conservation of flora and fauna in and around the Nature Reserve. In spite of these positive changes, monocropping of newly introduced cash crops, especially coffee, is expanding. This could become the dominant form of production, displacing the biodiversity-rich agroforestry systems and associated native crops, if the environmental and social benefits of these systems are not appreciated and supported.

22 Intensification and diversification of land use in the highlands of northern Thailand[1]

KANOK RERKASEM, CHARAL THONG-NGARN, CHAWALIT KORSAMPHAN, NARIT YIMYAM, BENJAVAN RERKASEM

NORTHERN THAILAND covers an area of approximately 171 000 km[2] and its 17 provinces are characterized by long mountain ridges and narrow valleys. Four major rivers, the Ping, Wang, Yom and Nan, flow southward and are the major tributaries of the country's biggest waterway, the Chao Phraya. The area shares borders with Laos and Myanmar, and contacts with neighbouring countries in the Mekong subregion can be traced back as far as 1050, the *Thai Era of Lan Na* (Penth, 2000). The population is now close to one million, made up of several main ethnic minority groups, including Karen[2], Hmong, Lahu, Lisu, Yao, Akha, H'tin, and other small minority groups such as Lua, Khamu, Shan and Yunnanese (Haw) Chinese (National Security Council and National Economic and Social Development Board, 1993; Department of Public Welfare, 1995). Virtually all of these people make a living by growing rice for subsistence and other crops for cash income, which in the days before the 1970s and 1980s included the opium poppy (*Papaver somniferum*). Originally, crop production activities were based on two broad groups of traditional shifting cultivation land-use systems, termed rotational and pioneer shifting cultivation (Kunstadter et. al., 1978; Grandstaff, 1980). Rotational shifting cultivators typically settled in one place, and grew crops in a rotation involving 1 year of cropping and 5–10 years of fallow. Pioneer shifting cultivators were migratory. Crops were grown on land cleared from mature forests, and the whole village would pick up and move to a new site after a few years of continuous cropping. Traditionally, opium was the major cash crop of the pioneer shifting cultivators. This system, which may or may not have been practised in its classic form, had been particularly severely abused for its destructive impact on biodiversity and the soil.

The land-use systems have undergone marked changes since 1960. In this chapter, we describe these changes in general, and provide examples from four villages to highlight some positive and negative aspects of the new systems.

The highlands in the context of national policy

Development efforts of the Thai government began in the highlands in the 1970s. With support from various international assistance schemes, they were directed at eradicating opium poppy cultivation. Central to these efforts were attempts to develop alternative cash crops. The first highland development master plan was initiated in 1983 with assistance from the United Nations Fund for Drug Abuse Control (UNFDAC). It targeted areas and groups involved in opium poppy cultivation (Office of the Narcotic Crops Control Board and UNFDAC, 1983). A second master plan of a similar nature followed in 1988. An important element of these master plans was the coordination of a large number of development activities initiated and supported by several international agencies and bilateral assistance agreements, and implemented as numerous 'highland development projects'.

These largely externally funded projects, which lasted until the late 1990s, helped to direct considerable public investment into the highlands in the form of road building, schools, health services and electrification, as well as the transfer of agricultural technology. Currently public investment for development comes from the Royal Thai Government. There has been a national master plan for highland community development, the environment and narcotic crop control for a period of five years from 1997 to 2001. Apart from this, the highlands receive public investment allocation on the same basis as the rest of the country. Support for development in the highlands is now at a much lower level than in the 30 years before 1990. The exceptions are a handful of villages that continue to receive substantial financial, technical and marketing assistance for their cash cropping through the Royal Project, which is partly funded privately through the Royal Project Foundation and partly publicly from a budget allocation to the Ministry of Agriculture and Cooperatives.

Establishment of permanent villages

Traditionally villages were highly mobile. Pioneer shifting cultivators moved in search of new forests after 5–10 years of continuous cropping. The villages of rotational shifting cultivators also split to establish new settlements when the population grew too large to be accommodated by the existing land. By the mid-1970s, however, movement had virtually stopped. Many Hmong, Lisu, Lahu and Akha villages became permanently settled in the 1960s or earlier. They acknowledged the increasing difficulty of finding new forests to clear. To settle, they frequently bought developed wet-rice land and the associated technology, including the irrigation system, from lowland Thai or Karen farmers. Apparently opium production was sufficiently productive to allow at least some highland farmers to accumulate enough wealth to buy irrigated wet-rice land and invest in commercial crop production.

The trend towards permanent settlement was reinforced by national policies instituted since the 1960s. Originally very few people in the mountains who belonged to any of the ethnic minority groups were recognized as citizens of Thailand.[3] Permanent settlement is still required as a first step towards official recognition and eventually to Thai citizenship. Citizenship has been granted to only about one-third of the population. Provision of health and education services, roads and electricity offered further incentives to settle permanently. There was also pressure through the national conservation and reforestation policy. Although the highlands had always been regarded by law as national property, they had until relatively recently been treated as a free good. In the past 40 years large areas of the highlands have been designated watershed areas, national parks, forest and wildlife reserves, with strict enforcement of conservation laws. All of these factors combined to make village movement and setting up of new settlements virtually impossible.

New cropping systems

New cropping systems have developed with permanent settlement. Thanks to strict enforcement of drug control laws, opium poppy cultivation has almost disappeared. To meet demand for home consumption by older addicts some small areas of cultivation remain, but these are well hidden. Wet-rice is grown by all ethnic groups, often in small highland valleys. Where dry season irrigation is available, rice may be followed by another crop, usually soybean or vegetables. Areas suitable for wet-rice are keenly sought after, but the amount of relatively flat land with sufficient water supply is limited. Cultivation on the slopes is still widespread, and much of it is on very steep gradients. Some land is cropped annually, some with two or three years fallow, and occasionally with the original full cycle of 5–10 years fallow. Upland rice, maize and various other food and cash crops are grown. The very short fallow periods, and sometimes lack of a fallow period, are associated with low yields and heavy weeding requirements. Farmers are reluctant to apply costly fertilizers and pesticides to subsistence crops, but do use them on high-value cash crops such as cabbages, coffee, tomato, potato, ginger, lettuce and flowers. Furthermore, all of the high-value crops that are grown in the dry season are irrigated by a system of sprinkler irrigation fed gravitationally from mountain streams and springs. These raise another set of problems.

Sustainability problems of cropping intensification

Improved national transportation, rising incomes in Bangkok and other cities, and a temperate environment, combine to create special opportunities for crop production in the highlands. The cooler climate provides an advantage for the production of temperate fruits, vegetables and flowers. During the monsoon season, vegetable production in the highlands has far fewer problems with insect damage than in the lowlands and there is better surface drainage on the slopes. Lychees are harvested much earlier and fetch very high prices. Research to find alternatives to opium and to evaluate new crop species and types began in the 1970s, and has continued with increasing commercial interests and initiatives. The two most recent additions are potato production to supply the fast-growing demand of manufacturers of potato products, and hybrid maize seed production. Commercial seed producers have discovered that the mountains provide the ideal conditions for isolation of populations to prevent unintended cross-fertilization between breeding lines.

All of the new cash crops are subject to wild price fluctuations. Downturns in prices threaten the food security of poor village families who have converted completely to cash cropping. The new crops require heavy fertilizer and pesticide applications. Intensive cultivation with a bare soil surface during the wet season contributes to soil erosion and has led to sedimentation behind dams and weirs and in paddy fields. There are also downstream hydrological consequences. Expansion of irrigated cropping in the highlands has been blamed for many mountain streams and springs running dry in the dry season. Conflicts have erupted between highland and lowland communities on these issues. In the Mae Taeng Irrigation Project of the Royal Irrigation Department, for example, the decline of dry season stream flow during the five months December to April from 1972 to 1991 led to an overall reduction of flow of 60.8 million m^3 over the 19 years, averaged at 3.2 million m^3 per year (Thailand Development Research Institute, 1995). In 1993 an ugly confrontation broke out between an upstream Hmong community in Paklouy village and lowland farmers in Chom Thong district of Chiang Mai valley who had their water supply for irrigation dry up (Benjasilaraks and Silaraks, 1999; Rakyuthitham, 2000). Few of the accusations and counter accusations are substantiated by actual measurements, and it is not certain how the problem of upstream and downstream conflict can be resolved in a near future.

Against this general background, the highlands are nevertheless a place of much diversity in both the environment and how farmers and communities respond to new challenges and opportunities. The case studies below, covering people with diverse ethnic backgrounds and contrasting traditional land-management systems, illustrate local innovation and adaptability. The villages are located in Chiang Mai and Mae Hong Son provinces (Figure 22.1). Two of these, Loh Pah Krai and Pah Poo Chom, are former opium-growing, pioneer shifting cultivator villages. The other two, Mae Rid Pagae and Tee Cha, are Karen, who are traditionally sedentary.

Loh Pah Krai – from opium to wet-rice and home gardens

This Lahu village had been a typical pioneer shifting cultivator village before the villagers settled at Mae Ai, north of Chiang Mai, in the mid-1960s. Within

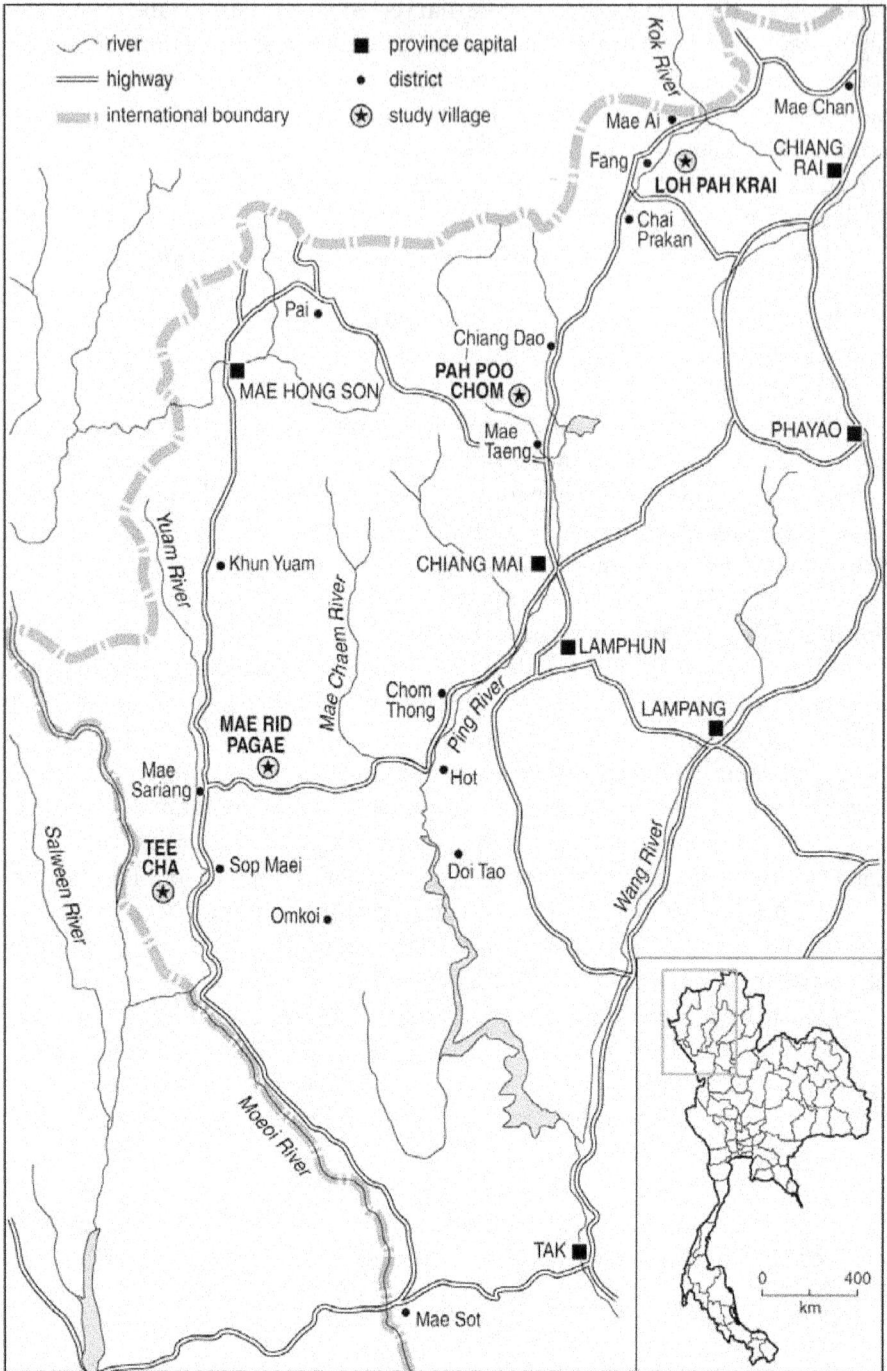

Figure 22.1   *Location of the case study villages in Chiang Mai and Mae Hong Son provinces, Thailand*

the life-time of some of the older members, the village had moved through and, in their own words, 'eaten up several forests' in Chiang Rai and Chiang Mai provinces. To settle permanently, the village bought a sizeable tract of irrigated rice land from some lowland Thais. This was not simply a transfer of land ownership. Along with the land, the Lahu farmers also secured the technological and management skills associated with wet-rice cultivation and the irrigation system. By the mid-1990s they had adopted a double crop system of rice and soybean in the wet-rice fields, and developed an effective scheme for sharing scarce dry season water with downstream villages as well as within the village.

When they first settled, the village was also growing some upland rice and maize on the slopes in short two or three year rotations. By the early 1990s most of the upland fields were cropped every year, often with double-cropping systems of maize or upland rice followed by a grain legume, such as soybean, payee (*Lablab purpureus*) or one of the *Vigna* species. A number of fields within ten minutes walk from the village houses had been developed into home gardens. Some 33 cultivated species were identified in one of the gardens (Table 22.1). Many of the species were grown for sale outside the village, but others provided a year-round supply of food, herbs, spices and animal feed. The food crops were readily shared within the village. Some species were incorporated into vegetative contour conservation strips, for example, lemongrass (*Cymbopogon citratus*), pigeonpea (*Cajanus cajun*) and cha-om (*Acacia pennata* subsp. *insuavis*).

Pah Poo Chom – many ways to biodiversity utilization and conservation

Pah Poo Chom is a Blue Hmong village situated in Mae Taeng watershed, north of Chiang Mai. The villagers settled on a mountain ridge at 940 m above sea level in 1963, and by 1970 most of the surrounding forests had been cleared and cropped with rice, maize and opium. According to the Tribal Research Centre of the Department of Public Welfare, the village was in a stage of extreme poverty, crop yields were low, and opium addicts accounted for 80 per cent of the adult male population and also included some women and even young children (Oughton, 1970; Oughton and Imong, 1970). Based on evidence from this village and many other highland villages, imminent collapse was predicted for highland cropping systems in the 1960s and 1970s (see Keen, 1972; Walker, 1975; Cooper, 1984). Since the early 1990s Pah Poo Chom has been transformed. Cash cropping has been adopted in a major way, but successful management of its biological diversity has also contributed to food security, income generation and conservation of biological diversity.

Table 22.1 **Crop species found in a home garden belonging to a farmer from Loh Pa Krai, Mae Ai District, Thailand**

Crop type	Common name	Scientific name
Introduced tree crops	Bamboo	*Dendrocalamus asper*
	Lychee	*Litchi chinensis*
	Santol	*Sandoricum koetjape*
	Mango	*Mangifera indica*
	Jackfruit	*Artocarpus heterophyllus*
	Tamarind	*Tamarindus indica*
New crops grown for cash	Adzuki bean	*Vigna angularis*
	Soybean	*Glycine max*
	Payee	*Lablab purpureus*
	Ginger	*Zingiber officinale*
	Green gram	*Vigna radiata*
Local plants	Cha-om	*Acacia pennata* subsp. *insuavis*
	Banana	*Musa sapientum*
	Papaya	*Carica papaya*
	Jujube	*Ziziphus jujuba*
	Upland rice	*Oryza sativa*
	Maize	*Zea mays*
	Sugarcane	*Saccharum officinarum*
	Sweet sorghum	*Sorghum vulgare*
	Ma Kua	*Solanum* spp.
	Chilli pepper	*Capsicum* spp.
	Pineapple	*Ananas comosus*
	Pumpkin	*Cucurbita moschata*
	Wax or white gourd	*Benincasa cerifera*
	Cowpea, several types	*Vigna unguiculata*
	Pigeonpea	*Cajanus cajun*
	Mustard green	*Brassica* spp.
	Taro	*Colocasia* spp.
	Pak Ped	*Vernonia silhetensis*
	Lemongrass	*Cymbopogon citratus*
	Sweet potato	*Ipomoea batatas*
	Tobacco	*Nicotiana tabacum*
	Sesame	*Sesamum indicum*

Source: Rerkasem et al. (1995)

Lychee trees and sprinkler-irrigated vegetables, principally cabbage, have become the main source of cash income. Villagers market most of their vegetables in Chiang Mai, carrying them in their own pick-up trucks. Agricultural land now accounts for only one-quarter of the village land. The balance is made up of natural forests, two parts conservation forest and one part utility forest dominated by bamboos, especially *Dendrocalamus* and *Bambusa* species (Figure 22.2). The largest number of species was found in the conservation forests, followed by the utility forests, the home gardens, agroforest edges between fields and lychee/vegetable intercrops (Table 22.2).[4] A large proportion of the species in each Land-use Stage are used. Harvesting bamboo shoots for sale is an important source of income especially for poorer villagers. Traditional crops that were part of the opium and upland rice swid-

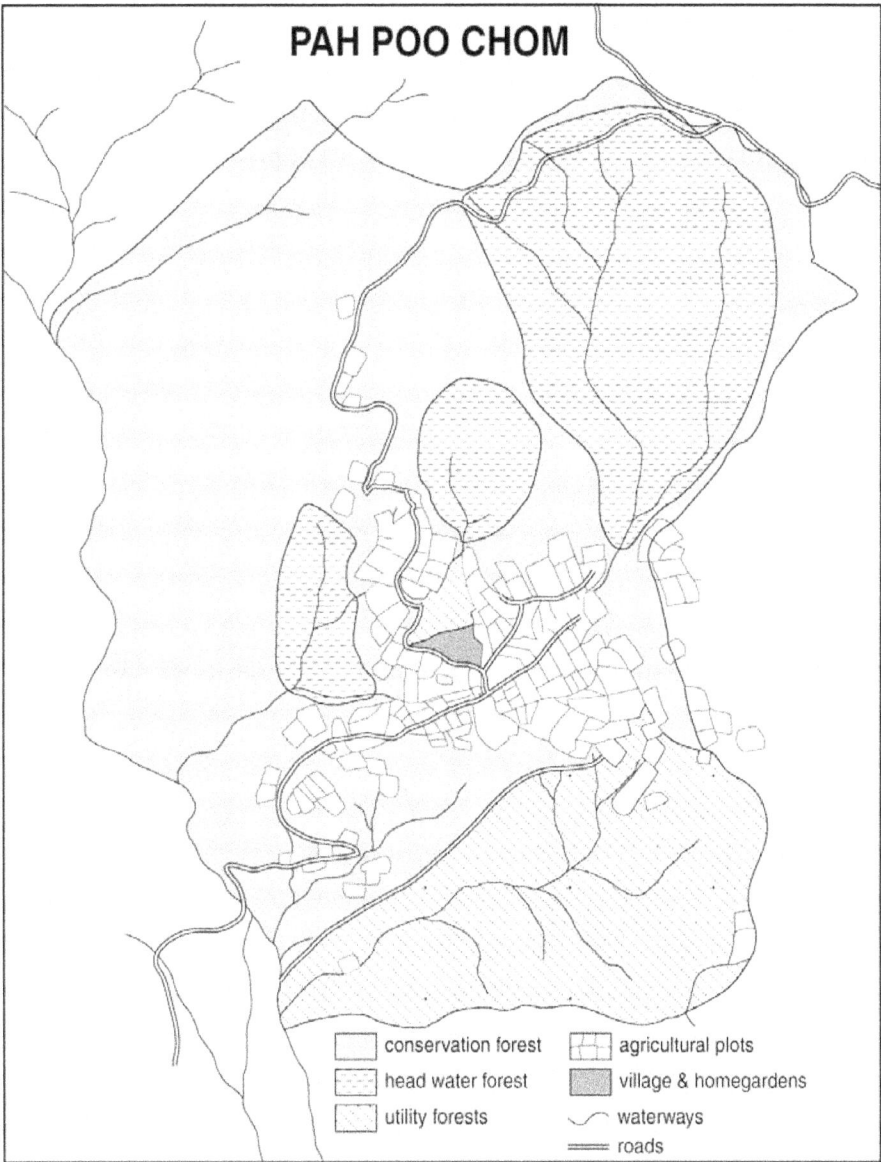

Figure 22.2   *Land-use map of Pah Poo Chom, Thailand*

dens have been conserved and incorporated into new cropping systems. The wild, semi-domesticated and traditional crops and vegetables from Pah Poo Chom (see, for example, Table 22.3) contribute significantly to village food security. They also find ready demand among the urban Hmong community around the Chiang Mai market. A traditional Hmong waxy or glutinous corn is now popular in the city market. Several clumps of a special bamboo are managed by one old man, who crafts them into the Can, a traditional Hmong musical instrument. Sold for 3000–4000 baht[5] each, several of these are made each year and are sometimes exported to Hmong communities in Laos.

Table 22.2   Number of plant species in various Land-use Stages and Field Types of Pah Poo Chom village, Thailand

Land-use Stage/Field Type	Number of Species*		
	Total	Used	% Used
Conservation forests (10 × 10 m)	152	133	87.5
Utility forests (10 × 10 m)	135	110	81.5
Bamboo dominant	89	82	92.1
Agroforest edges (total in the 3 sample plots)	89	68	76.4
Wild mango dominated (10 × 10 m)	33	27	81.8
Wetter area (10 × 30 m)	18	14	77.8
Patch close to village, near road (20 × 50 m)	63	46	73.0
Home gardens (total in two gardens)	68	57	83.8
Garden 1 (30 × 30 m)	45	38	84.4
Garden 2 (25 × 30 m.)	45	34	75.6
Lychee/vegetable intercrops (10 × 10 m)	34	12	35.3
Upper slopes	19	12	63.2
Lower slopes	12	5	41.7

*Numbers of species do not add up to the total in each category and grand total because some species occurred in more than one sample.
Source: Field Survey (1999)

Table 22.3   Useful plant species and their numbers in one semi-cultivated field (5 × 10 m) in Pah Poo Chom, Thailand

Species with common or local name			Number of plants in sample
Domesticated	*Zea mays*	Waxy corn (Kaopode in Hmong)	300
	Allium ascalonicum	Shallot	1150
	Brassica juncea	Leaf mustard (Pak-kahd in Hmong)	250
	Cucurbita moschata	Pumpkin	4
	Coriandrum sativum	Coriander	5
	Ipomoea batatas	Sweet potato	3
	Litchi chinensis	Lychee seedlings	1
Semi-domesticated	*Momordica* sp.	Wild bitter gourd	20
	Solanum torvum	Susumber	1
Wild herbaceous	*Crassocephalum crepidioides*	Lum Phasi	1
	Amaranthus viridis	Amaranth	1
	Asytasiella neesiana	Edible fern	11
	Phrynium capitatum	Tong Sard for food wrapping	9 clumps
	Musa acuminata	Wild banana	22 hills
	Zingiberaceae sp.	Kong: edible fruit, leaves for lining rice storage container, fibre for rope making	5 hills
	Total 15 species		1783

Source: Field Survey (2000)

Mae Rid Pagae – cash cropping improving food security

Mae Rid Pagae is a Skaw Karen village at 1200 m above sea level, some four hours by road from Chiang Mai. In the past, the limited irrigated wetland and 'sustainable' rotational shifting cultivation provided enough food for the population only in some years. There were sometimes bad years in which production fell short and many had to walk to the nearby town of Mae Sariang to seek work. As the population grew and the national conservation policy limited expansion of crop production into forest land, the problem of food security worsened. The luxury of adequate rice production with long fallow rotation in traditional shifting cultivation was impossible and farmers had to find viable alternatives to support their livelihood.

Cabbage production began in the early 1980s. Currently, visiting traders buy direct from farmers and truck the crop to Bangkok and beyond. However, it is not a monoculture of cabbage that has been adopted by Mae Rid Pagae. Instead, new cropping systems have evolved, incorporating traditional components and the cabbage. In the irrigated fields, cabbage is grown with irrigation in the dry season following the wet-season rice. On the slopes, upland rice and cabbage are grown in alternate years. The problem of sharply fluctuating prices has not eased, but rice yield has been greatly boosted by the incorporation of cabbage. This is probably due to the residual organic and inorganic fertilizers used in cabbage production, and the effect of clean-weeding of the cabbage in reducing weed infestation of rice. Farmers' reports of rice yield having doubled or tripled have since been confirmed in crop-cutting surveys.

The village also has the advantage of a well-structured soil that is less susceptible to erosion than many others. Vegetative contour strips have been adopted, incorporating weed species such as *Chromolaena odorata*, to check water flow down the slopes. However, farmers do acknowledge that they have begun to receive complaints from downstream villages about perceived water contamination with pesticides. In practice, the use of pesticides is limited to the dry season crops that are grown in areas with supplementary irrigation only.

Tee Cha – Pada fallow, a local innovation

In 1999 the population of Tee Cha numbered 148 in 41 households. The Pwo Karen village, established more than 200 years ago, is situated almost on the Myanmar border in Mae Hong Son province. It is one of a few villages where rotational shifting cultivation is still apparently sufficiently productive to meet food security needs (Figure 22.3). A good forest cover, with numerous uses and services, dominates land use. Being relatively isolated, lack of access to the market limits cash cropping. The cropping system in Tee Cha is

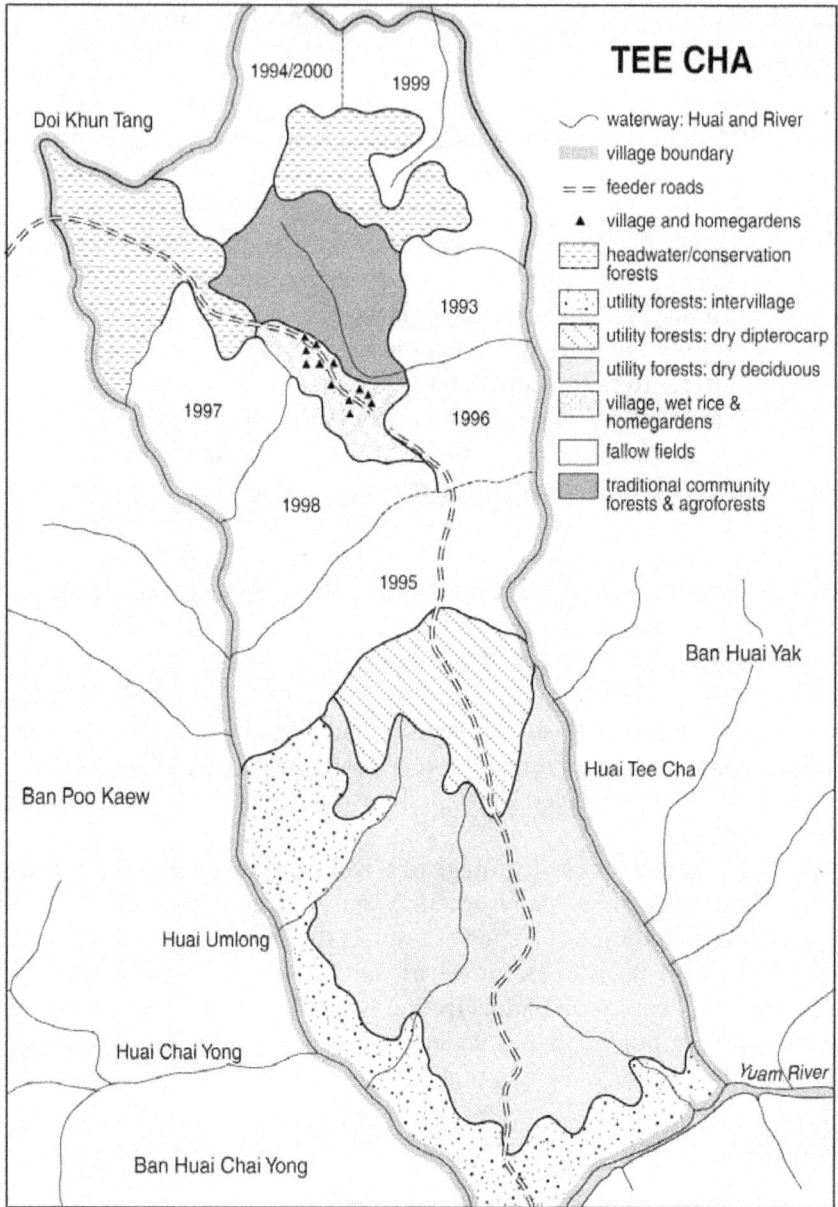

Figure 22.3 *Land-use map of Tee Cha, Thailand*

predominantly subsistence and is managed partly on a communal basis. Upland rice is grown in rotation with six years of fallow. The fallow management, mainly controlling fire and restricting the use of the regenerating forest, felling and burning the mature fallow, and the allocation of land-use rights, are communally organized. Managing the rice crop is an individual enterprise for each farm household, although there is much sharing and exchanging of germplasm and labour.

The seven-year cycle where land is cropped with upland rice every seventh year, is much shorter than the traditional ten-year or more cycle that used to be common in the region. Farmers attribute sustainability of the shorter cropping rotation to the dominance of Pada (*Macaranga denticulata* Muell. Arg.) during the fallow years. The fallow-enriching properties are recognized by other ethnic groups of northern Thailand, the H'tin and Khamu, who call it Teen Tao (Tong Tao in northern lowland Thai), and by forest ecologists (Whitmore, 1982). Pada is a small tree of the Euphorbiaceae family and it begins to flower and produce seed after three years. In the cropping year, Pada seedlings emerge as a thick carpet among the rice, presumably from a seed bank that accumulated during the previous three years. Farmers manage Pada in a number of ways. The seedlings are not considered weeds, so are not destroyed during hand weeding of the rice. However stands may be thinned if they become too dense and seedlings may be transplanted to areas with poor establishment.

The observations of farmers about the tree's value is corroborated by field observations. Preliminary measurements indicate that Pada trees play a major role in nutrient cycling of the cropping system. Nutrients such as nitrogen, phosphorus and potassium have been found to accumulate in the biomass of Pada-dominated fallow after the sixth year in much greater amounts than in Pada-less fallow after ten years (Zinke et al., 1978). A good stand of Pada, which reaches almost over the farmers' head by the time of rice harvest, is associated with an upland rice yield that is about twice the yield with few or no Pada. Attempts to establish Pada in neighbouring villages where it does not occur naturally, however, have so far been unsuccessful.[6]

Adapting to change

Since the 1960s land use in the highlands of Northern Thailand has undergone dramatic change due to external and internal pressures, which have included national conservation and highland development policies, population growth and rising expectations and aspirations of the villagers themselves. Intensive land use has replaced traditional shifting cultivation but some villagers, by using local innovations, have been better able to adjust their land-use systems to cope with the impact of change on food security and the environment.

Many other successful cropping systems and practices can be found in other villages throughout the mountains of northern Thailand. These observations belie the general belief that intensification of agriculture in the mountains is not sustainable and inevitably leads to yield decline and degradation. To understand how some farmers and villages in these difficult environments succeed, however, has required a holistic approach to the study of village land management that recognizes variations in time, space and the management units that exist in the agroecosystem. We conclude this chapter with three characteristic factors that have contributed to these farmers more successfully adapting to change.

First, the mountains are agroecosystems with great diversity. This diversity includes variability in:

○ the physical environment, for example, soil properties and microclimates
○ the social and economic context of the farming system
○ the local management capacity that may range from the different agronomic skills and ability to learn of the individual farmers, to the community's capacity to manage common resources or interact effectively with the market or the provincial and national government.

Second, there is a great diversity of plant genetic resources available in these mountain villages, including domesticated, semi-domesticated and wild species that are little known to outsiders, but that have been incorporated into successful new cropping systems.

Third, the innovations that have given rise to new cropping systems would not have materialized without the farmers' intimate knowledge of the specific sets of physical and socio-economic conditions defining their particular environment and the knowledge of the plant genetic resources available to them.

Innovations and knowledge originating from the outside can become useful only when they happen to fit local conditions. Modern agriculture and its various associated sciences have a great potential to benefit these mountain farmers. The challenge is to identify how they can be made relevant to local needs and conditions. Since it will never be possible to completely characterize the local variability, the best thing is to work closely with farmers.

23 Improvement of production and livelihood in the Fouta Djallon, République de Guinée

I. BOIRO, A. K. BARRY, A. DIALLO, S. FOFANA, F. MARA, A. BALDÉ, M. A. KANE AND O. BARRY

Introduction

THIS CHAPTER DESCRIBES the Fouta Djallon in Guinée, and in particular the Pita-Bantignel area in which the Guinée team has been working since 1994, and reviews of some successful enterprises established by the farming people in collaboration with PLEC scientists since 1997. It draws on a number of specialized and general reports written by the authors during the period 1996 to 2001, of which only one has hitherto been published. Anthropological work done some 60–80 km to the northwest of Pita, by Dupire (1970) and Derman (1973), has assisted interpretation of a highly complex set of social relationships which affect both the use of land, and what PLEC can achieve. The authors are grateful to Harold Brookfield for assistance in putting this material together.

The Fouta Djallon

The upland Fouta Djallon area in central Guinée lies at between 600 and 1500 m above sea level and is the source region of rivers draining south and west to the nearby Atlantic, north to the Gambia and Sénégal and east to the Niger (Figure 23.1, inset). With an area of 65 000 km², the plateau has 1.6 million people at densities ranging between 44 and 120 per km². Developed mainly on Palaeozoic sandstones and conglomerates, with massive dolerite intrusions, the dissected plateau is deeply weathered. Large areas have deep lateritic formations capped by ferruginous *bowal*, or duricrust. In the demonstration site area, all remaining forest is secondary and most of the landscape is covered in savanna. Rainfall of around 1500 mm falls mainly in the summer period from May to September. Most of the rest of the year is dry, and there have been as many as five rainless months at Pita in several years since 1960. The onset of the wet season is a period of heavy rains, giving rise to significant sheet erosion on land with little vegetative cover.

Soils

Soils are infertile, although harmattan dust from the Sahara adds some mineral fertility each dry season; heavier falls added much more during the arid periods of the Pleistocene. Soils are classified by the people, and by scientists, along a complex series of catenas in a dissected landscape. The duricrust (*bowal*) soils are extremely thin, and agriculturally useless. Where the duricrust is absent, the upland soils (*n'diare*) are sandy, with sandstone at shallow depth. The stony upper catena soils on slopes (*hansaghèré*) are the best agricultural soils, because they permit water percolation and retention, are supplied from uphill, have a good structure despite the large proportion of coarse gravels, and are less acid than other soils. The middle colluvial soils (*n'dantari*), though well drained, are sandy and acidic, while the clayey lower colluvial soils (*hollandé*) are both acid and hydromorphic. The alluvial (*dounkiré*) soils of the valley floors and terraces are waterlogged in the wet season, but hold sufficient water for dry-season cropping with manual irrigation (Barry et al., 1996). Along some of the streams are narrow belts of mainly degraded dry gallery forest.

The production system in its natural and social setting

The historical background

The Fouta Djallon has been occupied by people for well over 1000 years, and until the eighteenth century was occupied mainly by the agricultural Djallonké people. They were joined after the fourteenth century by pastoral Fulße (also known as Peulh, Peul, or – more internationally – Fulani), who developed a pattern of seasonal migration between summer pastures on the *bowal*, and dry-season pasture in the valleys on Djallonké land. Early in the eighteenth century there was a new and much larger invasion of Islamic Fulße, who conquered the region in a prolonged 'holy war'. Many Djallonké were expelled, but others, together with the earlier Fulße, were largely absorbed and converted to Islam. Fulße enslaved many of the Djallonké, and brought in much larger numbers of slaves through their wars of conquest.

Under colonial rule in the twentieth century, new slaves ceased to be available. Slavery was formally abolished, but the status of the slave population changed only into that of a serf class. Until 1949, they continued to be required to provide labour for their Fulße masters, and owned no land, receiving all they used from the Fulße who could dispossess them at will. But they had control over their own subsistence, and during colonial time, tensions arose as some of the serfs sought to claim land. Many used their freedom to leave the Fouta Djallon. Final abolition of servitude came only with land reform after independence in 1958, but inequality in status and access to land remained, and still remains.

Social and spatial organization of farming

The known agricultural history of the plateau dates from some time after the Fulße conquest. Initially, Fulße used their slaves and subjects for most agricultural and craft work, while themselves retaining control over cattle. Adoption of farming took place slowly. Over a period of between one and two centuries, the Fulße and their clients developed a distinctive infield–out-field system for management of their parsimonious environment. A pattern of settlement that has endured was established early, well recorded by Dupire (1970) in the Labé region. The largest domestic unit was a cluster of dispersed enclosures (*tapades*) containing both houses and fields. The enclosures form-ing these clusters were all similar in form, but some belonged to the dominant Fulße, including in each case the core *tapade* in which the mosque was built (*missidè*), while others might belong to lower orders among the Fulße and their subjects (*fulasso*), and still others were the dwellings and home farms of slaves (*runde*).[1] As population grew, new sets of clusters were established. Until quite recently new Fulße clusters might be established on the site of slave clusters, the slaves being relocated elsewhere. This was very much the pattern in the Bantignel area near Pita, which has long been one of the most densely-peopled parts of the Fouta Djallon (Richard-Molard, 1951). In the part of Bantignel covered by Figure 23.1, about one-fifth is occupied by the land of villages or *tapades*.

The group of villages comprising the administrative sectors of Missidè Héïré and Dar ès Salam, in which most PLEC work has been concentrated since 1997, display all the complexity of land-holding in the Fouta Djallon (Figure 23.1). Both were occupied in the eighteenth and early nineteenth centuries by Fulße who displaced or enslaved former inhabitants, now mostly assimilated into Fulße society. Founded by Koranic teachers who were invited to settle by earlier conquering Fulße, Missidè Héïré, and later Dar ès Salam, received land for *tapades* from these earlier landowning fam-ilies in the villages of Hindé and Bantignel Maounde, that lie outside the immediate area discussed here. The descendants of these original landown-ers continue to own the large areas in which external fields are made, renting them out but resisting all designs for permanent land transfer. In each cluster of *tapades*, the founding families continue to form the core groups. Missidè Héïré is the older; founded in the eighteenth century; Dar ès Salam became a principal village only after land was acquired for its *tapade* in the 1950s. But whereas Missidè Héïré is on *hansanghèré* soils, the greater part of Dar ès Salam and its neighbours are on the *n'dantari*.[2] Both sectors are some way from the permanent road system, and although there have been important improvements in the 1990s, they are poorly supplied with infrastructure. Missidè Héïré has only a single borehole for water, and no school or health clinic. The nearest health clinic is at Bantignel Centre, some 15 km away. Koranic education of boys is well developed in both com-munities, but only a few (about 13 per cent) are literate in French. As a community effort, a primary school was under construction in Dar ès Salam in 2000.

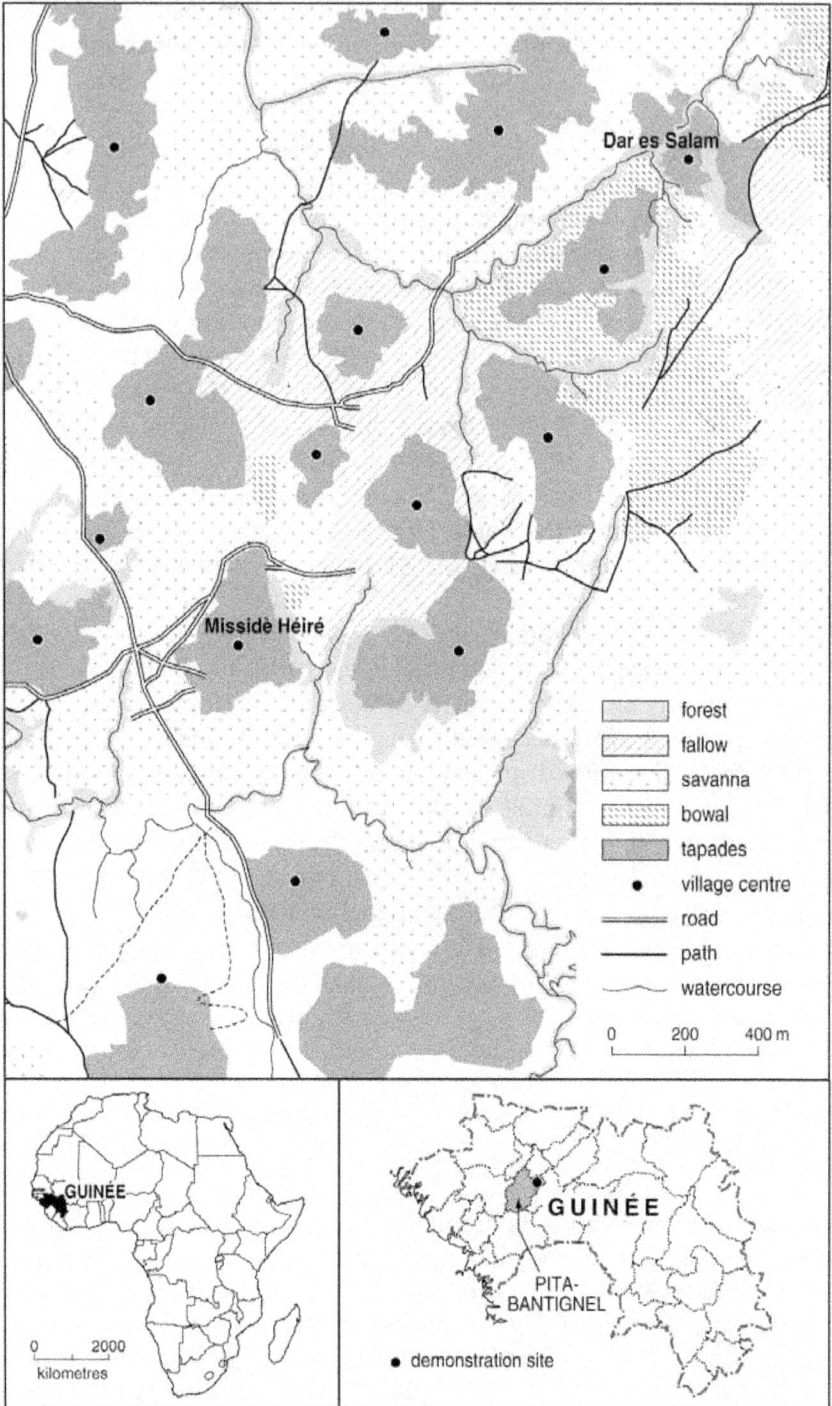

Figure 23.1   *Land-use in Bantignel region of the Fouta Djallon in Guinée*

The tapades

The *tapades*, which have much the same form whatever the original social class of their occupants, are at once the villages, the core of the system of land-holding and allocation, and the intensively cultivated infields. They contain the houses of landholders and their wives, usually with one concession (about 0.25 up to 0.5 ha) per wife. The wife 'owns' her concession while she remains married, but inheritance and transfer were and are only to male descendants. In the slave *tapades*, family or joint family residence was common. The family concessions, combining the allocations of different wives of the same husband where there are more than one (*sunturés*), are each enclosed by a fence, and form a network (Figure 23.2). Certainly from as long ago as the nineteenth century, the *sunturés* were intensively cultivated with large imports of manure and recycling of vegetable and other wastes. The manure was supplied by the small n'dama cattle of the Fulße, as well as goats and sheep, the proportion of which has increased through time. The livestock are pastured on the open lands outside the *tapades*, as well as being stall-fed within them. Today, cattle are normally corralled at night just inside the boundary fences, while goats and sheep have pens within the enclosure that are approached by narrow runways from the boundary fences.

Sunturés are areas of manufactured soil created within a manufactured micro-environment. Fencing is a male task, although formerly the women had to collect the wood from a distance. In old-established settlements, live hedges – which require less maintenance and also provide food for cattle and mulch for the ground – largely take the place of the old dry-wood fences, now also replaced by wire fences. Men also plant and own fruit trees within the *tapades*. Women and children do almost all other work. Before the rains have begun, and after taro and cassava remaining from the previous year are lifted from the ground, manure is collected, dried, spread on the ground and hoed in. Taro is immediately replanted, and as soon as the rains begin maize is planted, followed by cassava, groundnuts and sweet potatoes. A number of other plants, some of them recent introductions, are also planted, including tomatoes, courge (*Cucurbita pepo*), niébé (cowpea, *Vigna unguicalata*), pea aubergine (*Solanum torvum*), chile (*Capsicum frutescens*), gombo (*Hibiscus esculenta*), yam (*Dioscorea alata*), haricot beans (*Phaseolus vulgaris*) and potatoes. Then the whole surface between the growing crops is covered in leaves, both from within the *tapade* and outside, mostly as mulch, but sometimes composted together with animal manure. Weeding and the further spreading of ash, manure and domestic refuse on the land continue through the year. Small quantities of chemical fertilizer, mostly provided by the authorities or aid projects, are applied on a precision basis to the plant sites. Maize is harvested at the end of the wet season, and the *tapades* yield mainly root crops and fruit until the following wet season. In the *tapade* at Missidè

Héïré, yields were measured on 6.5 ha and it was found that maize yielded up to 7 t/ha, cassava 21 t/ha, sweet potatoes 19 t/ha and groundnuts about 8 t/ha.

Depending on fertilizer input, it takes ten years or longer before a *sunturé* achieves its full production, with input of nutrients fully replacing the harvest. When maturity has been achieved, the pH within the *tapade* is up to 1.5 points higher than in the open land outside. Soil structure and chemical content, except phosphorus, reach satisfactory levels. Although heavy application of manure can override the natural differences, *tapades* on *hansaghèré* soils, such as those of Missidè Héïré, often do better than on *n'dantari* soils. The majority of *tapades* – and especially those of the Fulße as opposed to the former serfs – have been made on the former. The *n'dantari* lands are mainly devoted to rotational use under the external fields.

The external fields

In a characteristic infield–outfield manner, the fields outside the *tapades* are deprived of nutrients in favour of the infields. Formerly burned annually, and still cleared with the aid of fire, these fields are cultivated from 3–5 or 7 years, then fallowed for at least six or seven years, and sometimes longer. They are not allocated to individuals as are the enclosures within the tapades, nor are they collectively owned as are the 3 ha of forested land and 12.88 ha of shrub savanna, reserved from use outside the village of Missidè Héïré (Figure 23.2). Even now, the large external areas comprising two-thirds of all the land, remain owned by the descendants of those who held them at the time the settlement pattern was established. These are the 'nobles', or highest class, among the Fulße of Hindé and Bantignel Maounde, situated some kilometres away. At Missidè Héïré, the core families own almost all the land within their *tapade*. A good deal of this land has been bought and sold in modern times, but the holders of land within the *tapade* own none of the surrounding land, and their external fields are from 2 km to as much as 10 km away. External land may sometimes be rented for a single payment covering the five years or so of cultivation, but most land is rented annually on sharecropping tenancies at an annual fee of one-tenth of the crop (the *farilla*).[3] Because of land shortage in this densely-populated area, individual plots in the external fields are seldom larger than 1 ha.

Work in the outlying fields is also done mainly by the women and children, except for some of the heavier hoeing work and use of an ox-plough where this is possible. Clearing and burning are dry-season jobs, and planting begins as early in the wet season as the prior demands of establishing the maize and other crops in the *tapades* will allow. Often, three weedings are necessary, and the women do all these. The undemanding fonio (*Digitaria exilis*), grown almost entirely in these outfields, is the basic grain crop in Fulße subsistence agriculture.

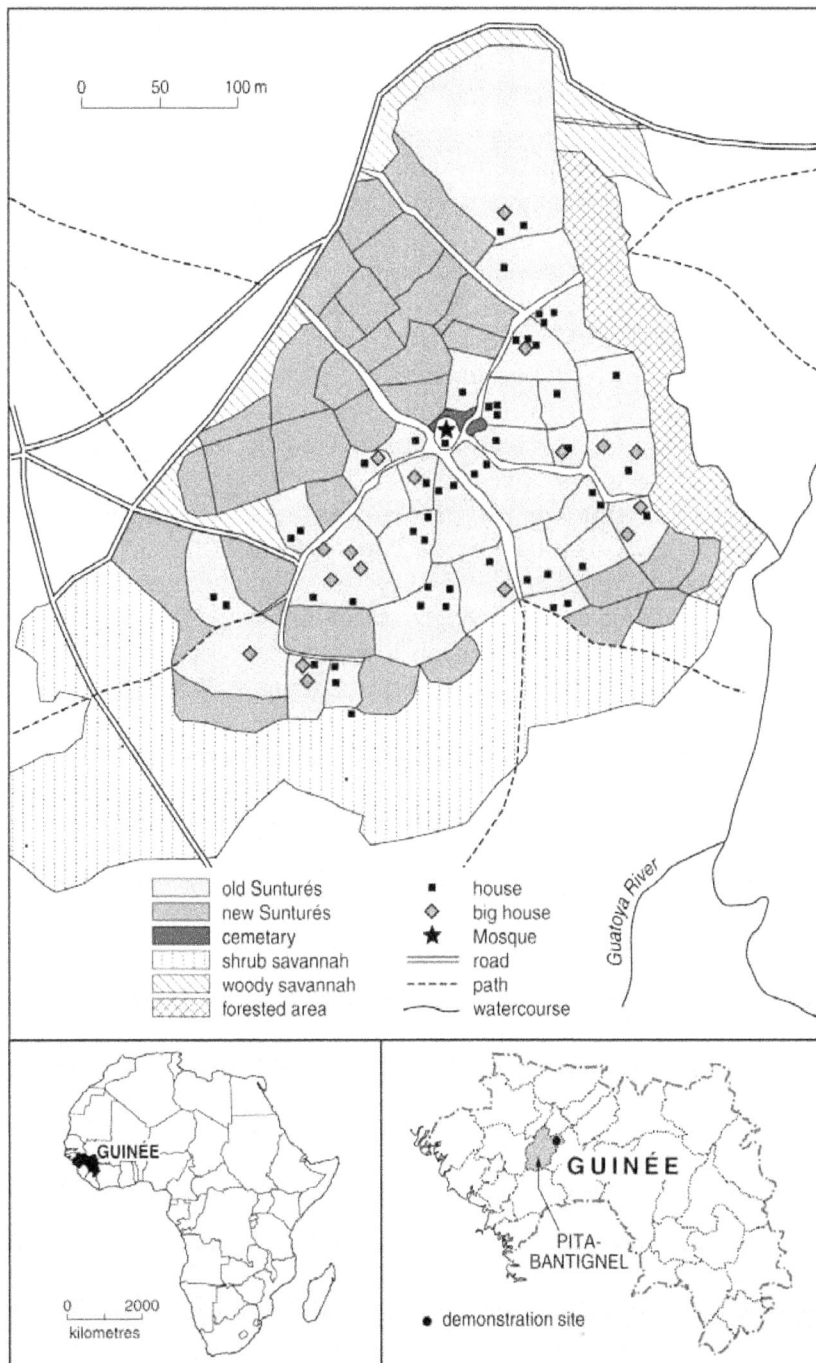

Figure 23.2   *Land-use in Missidè Héïré, Fouta Djallon in Guinée*

The type of cultivation and fallow cover depend on the soil. The stony but more fertile *hansaghèrè* soils permit longer periods of cultivation, and support a low woody fallow. Those on steeper slopes can be cultivated only by hand. The *hansaghèrè* soils will support four crops, upland rice, then groundnuts and two crops of fonio, before being fallowed. It is only on these soils that upland rice, formerly a much more important crop on the Fouta Djallon than it is today, is still sown. Cultivation on the sloping and more stony areas is considerably reduced in modern times, because of the large out-migration of men discussed below.

Larger field areas are developed on the *n'dantari* soils. Their stone-free constitution permits the use of the ox-drawn plough, an introduction from the colonial period. These soils will, however, support only two or three crops, perhaps fonio, followed by groundnuts then fonio, or perhaps only fonio. There are only limited areas of nitrogen-fixing groundnuts. On the smaller areas of hydromorphic (*hollandé*) soils, lowland rice is possible as the first and even third crop in a four-year cycle.

Since livestock pasture freely by day outside the *tapades*, the cropped areas in the external fields must be fenced. Cattle can break through weak fences, which last only one year before they have to be rebuilt; previous fences are employed as a source of increasingly scarce firewood. The productivity of most of the external fields is extremely low, and is declining due to heavy use with insufficient fallow, collection of livestock dung from them for use on the more intensively-cultivated land, and their liability to sheet erosion, especially on the *n'dantari* soils.

The alluvial areas along the streams

These small areas have year-round use, which their clayey *dounkiré* soils will support. They are cultivated for lowland rice and maize in the wet season, and in the dry season, with daily hand-irrigation, they nowadays support a variety of vegetable crops for sale in the nearby weekly markets. Some are fertilized with urea and other compounds bought in those markets. Organic manures are also applied, especially on seedbeds. Dry-season farming in the valleys is less labour-intensive than in the *tapades*, but not by a wide margin.

The modern constraints on the social and production systems

Although the *tapades* provide about half the food required for subsistence, and continue to do so year-in, year-out for long periods, the possibilities for expansion are constrained. At Missidè Héïré, rather more than half the area of the 62 *sunturés* is old and established, covering 15.65 ha, but only 26 of these *sunturés*, covering 10.5 ha, contain dwelling houses and receive full support of domestic refuse for fertility. The remaining 11.51 ha is more newly established, and has neither the high manufactured fertility of the old *sun-turés*, nor their year-round production. Dar ès Salam is, as a whole, more recent, but more manure is applied, and productivity is therefore better than at Missidè Héïré. The problem of productivity is in part shortage of labour, but most important is the shortage of manure. While every family has small livestock, only some have cattle, although there may be more than 150 cattle at Missidè Héïré. Because these are pastured freely outside the *tapades* during the day, a good deal of their manure is lost, and there is insufficient to supply the newer *sunturés* with as much as they need.

The importance of emigration

Unlike what happened in the Fulße regions of northern Nigeria, the colonial period did not bring any successful cash cropping to the Fouta Djallon, except at a local level where fruit, onions, maize, fonio, taro and tomatoes are sold in the small weekly markets. Colonialism did, however, bring tax, which had to be earned by the use of labour, and this included the labour of the Fulße as well as the former slaves. As the demand for money incomes grew, seasonal migration to the groundnut areas of Sénégal was already well established before the Second World War. In subsequent years it has also been augmented by migration to the towns of Guinea, Côte d'Ivoire, and Sierra Leone as well as Sénégal. In consequence, women substantially outnumber men between the ages of 15 and 49. The emigrants do, however, invest at home, especially in improved housing, and a considerable number return to settle toward the end of their working lives.

Largely because of emigration, not all the *sunturés* are now occupied. At Missidè Héïré, as many as one-third are currently wholly or partly idle, or are cropped without fertility maintenance. Some of the *sunturés* are used on loan with payment of rent by farmers wishing to expand their production. Others are bought from former owners, but prices have risen sharply in recent years. Women whose husbands are away may find themselves with insufficient land for their needs.

PLEC initiatives in the villages

Within a seriously degraded region, the small pockets of intensive agriculture have for some time failed to support an economy which yields any significant surplus to its people. The great inequalities in access to resources are only gradually being overcome by national legislation and infrastructure investment. During the late 1980s and the 1990s, however, a number of externally-sponsored initiatives took place in the general region of Pita, and these included some reforestation, initiatives in compost making, and some useful agronomic improvements, including row-planting. Two leading farmers from Missidè Héïré and Dar ès Salam were among those who learned new ways through these interventions. When the PLEC team arrived looking for leaders who might organize people for conservation with development, these two farmers – who are among the authors of this chapter – collaborated actively with the project scientists in a whole series of new activities. These have included reforestation and the general replacement of dry-wood fences by live fences strengthened by wire, and the development of potatoes, onions and other vegetables as dry-season cash crops as an important source of revenue to the women. Other innovations have included the introduction of coffee, extension of agroforestry, and modernization of compost production based on the construction of cowsheds and improved composting technology. Also important have been revitalization of dyeing of cotton clothing, soap making, improvement of fonio cultivation, and the training of more than 25 women farmers to read and write. By demand, several of these activities have been extended to nearby villages outside the initial group, and close relations have been developed with the regional authorities, a local NGO, two aid projects, and the agricultural research station at Bareng, near Pita. Two of these initiatives are discussed in greater detail below.

Cattle manure and compost making

Making compost using scientifically advanced methods was an early initiative in 1997 at Missidè Héïré, where the leading farmer introduced technology he had learned from specialists brought from Bareng, and had subsequently perfected (Fofana et al., 1998). In collaboration with an NGO, good cowsheds were constructed first at one village, then two more, to improve both the supply and quality of manure. This improvement has been extended to further villages in 2000 and 2001. The compost was initially used to fertilize potatoes grown in rotation with maize, and has more recently been applied also to onions and fonio, as well as to dry-season vegetable crops grown near the houses and in the shallow valleys using irrigation. Texture and chemical properties of the soil were notably improved, approximately doubling the organic matter content at Missidè Héïré and by 1.5 times at Dar ès Salam.

Table 23.1 shows that potato and maize yields showed notable improvement, and this became dramatic when small quantities of a chemical fertilizer were also added. Most striking was the speed with which improved fertility and productivity were achieved by the use of compost, a considerable acceleration on the results obtained by traditional methods in the old *tapades*. There continued to be problems with seasonality, involving waterlogging of the crop at a third site, and crop losses due to spoiling. Moreover, the potato crop peaked at a season of low prices, leading in 2000 to experiments with changes in the farming calendar to take greater comparative advantage of the higher productivity.

Table 23.1   Effect on crop yield of composting and chemical fertilizer application at Missidè Héïré and Dar ès Salam, 1998–2000

Sites	Years	Crops	Treatments	Yields (t/ha)
Missidè Héïré	Nov 1998	Potatoes	Control	2.0
			Compost	7.2
			Compost + triple 17	15.4
	June 1999	Maize	Control	0.75
			Compost	1.2
			Compost + triple 17	1.86
	June 2000	Maize	Compost	1.55
			Compost + triple 17	2.36
Dar ès Salam	Nov 1998	Potatoes	Control	3.5
			Compost	8.12
			Compost + triple 17	16.45
	June 1999	Maize	Control	0.57
			Compost	1.3
			Compost + triple 17	1.94
	June 2000	Maize	Compost	1.44
			Compost + triple 17	2.5
	June 1999 (extension)	Potatoes	Compost + triple 17	17.23
	June 2000 (extension)	Potatoes	Compost + triple 17	16.72

Activities specific to the women: dyeing and soap-making

Most of the work on potatoes and maize was done by women farmers, as also was the increased cultivation and sale of groundnuts, onions and fonio. However, the making of compost requires male labour, and in the absence of wheeled vehicles, transporting compost over any distance is very demanding of physical endurance. Revitalization of dyeing cotton cloth provided a less arduous enterprise specifically for the women. The use of a range of vegetable dyes is an old activity in the Fouta Djallon, formerly limited to a single occupational 'caste' group among the subjects of the Fulße royal houses. It died out in the colonial period because of competition from imports, but since the 1920s there have been several efforts to revive it by Fulße women using the strong blue dyes obtained from wild plants of the Fabaceae family, *Lonchocarpus cyanescens* and *Indigofera tinctoria*. In the 1980s, a woman from Missidè Héïré began a new effort, in which she has been materially supported by PLEC since 1997. A group of ten women was formed, making a very good and visually striking product, especially when using local fabric, the manufacture of which has also increased. Wild sources of the plants are close to the village but are limited, and by 2000 were already becoming threatened by increased extraction. PLEC began to support replanting. The dyeing enterprise has been remarkably successful, and in 2000 in Missidè Héïré it became a substantially larger income earner than the potatoes, onions and tomatoes. Moreover, all this income came to the women.

Soap-making was initiated at Dar ès Salam using wild plants, in particular *Jatropha curcas* and *Carapa procera*. In 2000 it was extended to a village outside the original group. While less profitable than dyeing, it provides a useful product for sale in the local weekly and daily markets.

Conclusion

All these improvements have taken place in and around the *tapade* villages, the furthest extension being reforestation in adjacent common land and in the headwaters of small streams which mostly run dry in the rainless months. Nurseries of species useful for food or income, medicinal purposes, firewood or simply for forest recovery, have been set up outside Missidè Héïré and one of the communities in Dar ès Salam sector. There have also been efforts to control drainage on to the village fields, and to manage soil erosion in the immediate vicinity. But not much else has been possible at landscape level, above all in the two-thirds of useful land comprising the external fields and their fallows. Yet these are the main areas in which land degradation continues to take place. The conditions of tenure of the external fields still provide no long-term security for the cultivator, and improvements such as tree planting are specifically prohibited by the owners. None of the national land reform measures, not even the major one of 1992, has yet had a significant effect on the more intractable land tenure problems of the Fouta Djallon.

Formation of human capacity for improved resource management, and ability to negotiate written conditions of tenure, are basic to what might be achieved in the future. Since 1998, PLEC has been very successful in organizing village groupings for collaborative work and planning. There are over 47 actively participating farmers in the Bantignel area, all but four of them women, and the number grows each year. In 2000, it began to include participants from two of the oldest villages in the area, Hindé and Bantignel Tokossèrè. Hindé is the place of residence of some of the greatest landowners. Two of PLEC's most valued interventions were extended to Hindé in 2000. Writing of the social constraints in the region in an undated report, written in the late 1990s, S. Fofana and P.O. Camara stressed the great need to develop a culture involving sale of land and long-term contracted tenure in the region. The informalities of unwritten agreements are successful where the demand for land is low in relation to population, as at the lowland site at Moussaya, on the upper Niger, also developed by PLEC. However, in the high density Fouta Djallon, with its complex history of inequalities that even now affect the use of land, more radical improvements in resource management at landscape level are required. They demand formal conditions that will protect the rights of owners and tenants and at the same time permit collaboration for the conservation and rehabilitation of natural resources.

24 Traditional forms of conserving biodiversity within agriculture: their changing character in Ghana

EDWIN A. GYASI

Introduction

A MAJOR GLOBAL challenge, especially in developing countries, is to increase and secure food production for a growing population while, at the same time, conserving the natural diversity of crops, livestock, trees and other life forms in their natural state (United Nations Environment Programme, 1992, 1994). Many of the systems of farming developed over the years that meet this challenge are diverse and all exhibit a high degree of agrodiversity (Brookfield et al., 1999). They are founded on farmers' solid understanding of local conditions (Amanor, 1995; van den Breemer et al., 1995).

Biodiverse agricultural systems in Ghana are under threat. This chapter describes traditional systems of conserving biodiversity within agriculture in the semi-humid southern sector of Ghana's transitional forest-savanna ecozone (Figure 24.1). It discusses their development and changing characteristics, their relationship to food security, and identifies the forces threatening them. Research was carried out from 1993 onwards at three sites; Gyamfiase-Adenya, Sekesua-Osonson and Amanase-Whanabenya, within surrounding tracts that each cover an area of about 100 km². They have been developed as sites for demonstration of agrodiversity conservation and sustainable management of biophysical resources.

Historical overview

According to Okigbo (1998), farming systems in Africa are in most cases in a transition from traditional to modern systems with local adaptations (See also Benneh, 1972). Agriculture in the forest-savanna ecozone fits this model. Before about 1850, most of the zone consisted of sparsely inhabited old high forest land owned largely by Akyem people. Elsewhere, was a low-impact economy, including the hill areas inhabited by Akuapem people. The Akuapem already supplied food, palm oil and bush meat to nearby coastal settlements, but the principal activities were hunting, gathering and food cropping based upon agroforestry and bush fallow or land rotation.

Historically, agrodiversity, and by implication food security, were enhanced by the introduction and integration of additional crops through trading and

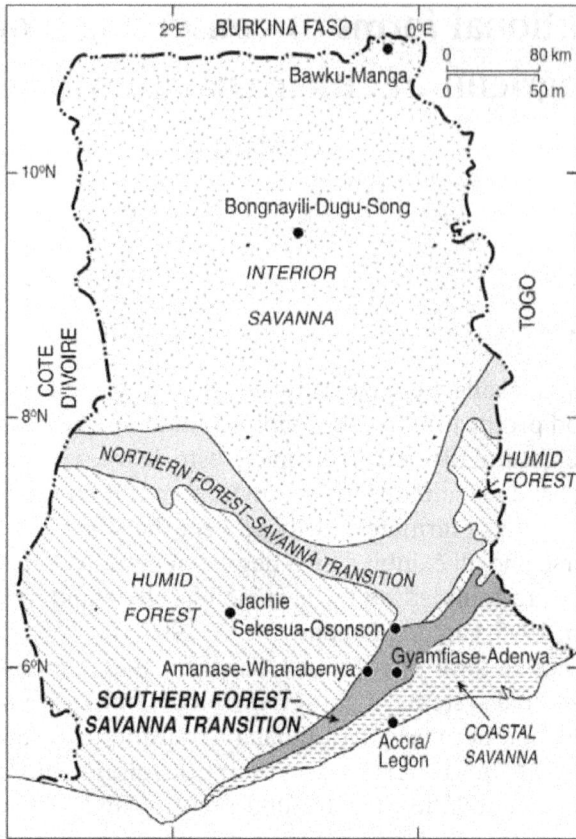

Figure 24.1   *The major agroecological zones and demonstration sites in Ghana*

other interaction between Ghana (then called the Gold Coast) and the external world. With reference to the period 1700–1850, when European traders maintained a series of 'castles' along the coast, Dickson writes:

> Almost every European castle owned a 'kitchen garden', not necessarily close to the castle, in which both local and temperate food crops were cultivated. Some of the gardens were fairly large and well laid out . . . They served also as experimental agricultural stations in which exotic and local crops were cultivated under careful management (Dickson, 1969: 75).

Crops introduced included coffee (*Coffea*) of African origin, and polyploid cotton (*Gossypium*), which probably reached West Africa from the Americas (Van Royen, 1954; Dickson, 1969). Crops of Asian origin included Asian rice (*Oryza sativa*), taro (*Colocasia*)[1], water yam (*Dioscorea alata*), ginger, peas, oranges, grapefruit, mango, coconut, plantain and banana, eggplant and sugarcane. Maize, cassava (manioc), sweet potato, common bean (*Phaseolus vulgaris*), groundnuts (*Arachis hypogea*), lima bean, cocoa, avocado, guava and cashew came from the Americas (Okigbo, 1998). The impact on the agroecology and the political economy of cocoa and cassava was particularly profound.

From about 1850, pressure on the sparsely inhabited forest areas increased substantially. This was a result of in-migration by Krobo and Akuapem farmers searching for more land, initially for export-oriented production of palm oil and kernels from both wild and newly cultivated *Elaeis guineensis* palms. The oil-palm expansion lasted until about 1900, when large-scale production of cocoa began. Basel missionaries, the British colonial administration, and Tetteh Quashie, a Ghanaian migrant who returned from Fernando Po, all were reported to have introduced cocoa into the Akuapem district in the second half of the nineteenth century. Cocoa gradually superseded oil palm as the primary source of export earnings. Cultivation of both export crops was predominantly by small-scale farmers and closely imitated the forest ecosystem with integration of food crops under and among trees and shrubs. Modernization included more systematic planting by smallholders and the introduction of the plantation system.

From the 1930s cocoa was devastated by disease, most especially the swollen shoot virus, but also by the mirid bug (or *akate* in Akan) associated with fungal disease, and planting in unfavourable soil conditions (Moss, 1969). Subsequently there was a shift to staple food crop production based on traditional, as well as modern management practices including the use of the tractor (as opposed to the cutlass and other manual tools for land preparation), monoculture, use of agrochemicals and hybrid seeds, and row planting. Recently there has been an increased use of leguminous crops (Gyasi and Uitto, 1997).

Agroforestry systems

Agroforestry is an age-old practice in tropical Africa. It involves cultivation of field crops among trees deliberately planted or left when clearing the land. The two agroforestry types in the study area were a type located adjacent to the home, which the author and Lewis Enu-Kwesi termed 'home garden agroforestry', and a type located away from the home in outfields, which we termed 'non-home garden agroforestry'. Home garden agroforestry invariably occupies the land more or less permanently, whereas the non-home garden agroforestry does not.

Non-home garden agroforestry

Under low population density, cropping among trees left in farms within easy walking distance of villages and towns was the predominant form of farming in the era before British colonization in the nineteenth century. The cropping system was based on bush fallowing whereby the land was farmed on a rotational basis around a fixed or permanent settlement. Sometimes exhausted croplands together with the settlements were completely abandoned in favour of new locations. Low population densities permitted long or indefinite fallow periods that allowed regeneration of soil fertility (Benneh, 1972).

Initially this agroforestry system was centred on African species of root and tuber crops, notably yam (*Dioscorea* spp.). Trees among which crops were interplanted included those difficult to fell, those perceived as sacred or performing unique ecological functions, and particularly those having medicinal or economic value, notably oil palms, kola nut (*Cola acuminata*) and timber species. The system featured:

- low-impact minimal tillage involving parsimonious use of fire, the cutlass and the hoe
- reduced risk of soil erosion under the tree canopy
- more effective use of soil nutrients by planting crops with different requirements
- ecological stability by combining crops with trees
- reduced risk of complete crop failure
- a more balanced diet through the diversity of crops.

The agroforestry system still features a diversity of trees. Table 24.1 shows the species of trees maintained on farms and their uses at Gyamfiase-Adenya during surveys since 1994.

Table 24.1   Trees recorded as conserved on farms at Gyamfiase-Adenya since 1994

Twi name	Botanical name	Use
Odwen	*Baphia nitida*	Medicinal, fuelwood
Akakapenpen	*Rauvolfia vomitoria*	Medicinal, fuelwood
Onyankyren	*Ficus exasperata*	Other
Osena/Yooye	*Dialium guineense*	Food, fuelwood and other
Ankye	*Blighia sapida*	Food, fuelwood and other
Abrewa aninsu	*Hoslundia opposita*	Medicinal, fuelwood, other
Osisriw	*Newbouldia laevis*	Medicinal, fuelwood
	Bryophyllum pinnatum	Medicinal, fuelwood
Pepea	*Phyllanthus discoideus*	Fuel
Emire	*Terminalia ivorensis*	Timber
Ofosow		Fuelwood
Osese	*Holarrhena floribunda*	Food, carving
Owudifo akete	*Anthocleista vogelii*	Medicinal
Okronoo	*Bombax buonopozense*	Medicinal
Awonwee	*Olax subscorpioidea*	Medicinal
Opanpan		Fuelwood
Nyamedua	*Alstonia boonei*	Medicinal
Agyama	*Alchornea cordifolia*	Medicinal
Bronyadua	*Morinda lucida*	Medicinal
Adurubrafo	*Mareya micrantha*	Medicinal
Odwendwenaa	*Lecaniodiscus cupanioides*	Other
Atwere		Other
Atuaa	*Spondias mombin*	Medicinal
Kyenkyen	*Antiaris toxicaria*	Food, other
Odum	*Milicia excelsa*	Timber

Cocoa was introduced into the agroforestry system and its spread was largely by migrant farmers through the company or group purchase system of land acquisition called the *huza* system. The migrants were mostly Akuapem people from the Akuapem hills within the forest-savanna zone and Krobo people from the southern fringes of the transitional ecozone that then comprised mainly thick forest. As they moved into new agricultural frontier areas, Akuapem people introduced their traditional root crops of yams, taro and cassava. These crops were grown together with the cocoa among naturally occurring trees. They were also grown in association with maize and other food crops, and sometimes among oil palms and coffee in separate patches of land.

The Krobo farmers also introduced cocoa among the root crops that in part had been introduced to them from the neighbouring Ewe and Akan areas. They grew white guinea yam, or *hie* in Krobo language (*Dioscorea rotundata*), yellow yam or *kani* (*Dioscorea cayenensis*), water yam or *alamoa* (*Dioscorea alata*), taro and cassava (Huber, 1993). The Krobo had virtually abandoned millet (*Panicum*), their traditional staple, in favour of these root crops. They, like the Akuapem, commonly grew the root crops in separate farms, frequently among oil palms which were highly prized by the Krobo (Field, 1943; Hill, 1963; Dickson, 1969; Kwamena-Poh, 1973; Huber, 1993; Amanor, 1994; Gyasi, 1992, 1994, 1996b; Gyasi and Uitto, 1997).

The mixed cropping improved food security by increasing the diversity of crops with different maturation periods and responses to environmental stress. However, intercropping with cocoa disadvantaged food crops. The cocoa canopy expanded and shaded the crops as the trees matured. This disadvantage was offset to some extent by growing shade-loving food crops, notably some yam types.

One of the strategies developed by farmers to counter the effects of swollen shoot disease in cocoa was specialization in food crop production for the expanding urban markets. Cassava was central to this agricultural transition because it was tolerant of the increasingly acidic soils and drier climate that disfavoured cocoa. It is high-yielding, affordable and able to meet the food requirements of a population expanding at more than 3 per cent per annum (African Development Bank, 1995; Christiaensen et al., 1995; Asante, 2002). Fourteen cassava types were identified by their local names in surveys since 1994 in the three study sites. Some of the varieties were introduced only recently from neighbouring countries by immigrants and returned Ghanaian migrants, including the variety agege (or biafra) from Nigeria, and kable from Togo. This highlights the continuing importance of migration in changing the character of agrodiversity.

Less dramatic, but nonetheless significant, has been the increased emphasis on maize, and also vegetables and legumes whose cultivation is gradually involving hand irrigation, manuring, agrochemicals and other forms of intensification. The forest-savanna ecozone has consolidated its role as a

major producer of food crops, notably cassava, maize and vegetables for the
nearby coastal urban areas and other settlements in Ghana (Gyasi, 1997; see
also Nabila, 1997). Yet there are widespread claims of deteriorated food security in the study sites. In a 1995 PLEC sample survey, 77 per cent of farmers
reported that agricultural yields had declined, while 34 per cent reported an
inability to feed their households adequately solely from their farm produce.

Home garden agroforestry

The home garden agroforestry system has endured for many years in Ghana,
and it also occurs throughout the forest-savanna zone. It is best developed in
Sekesua-Osonson, settled by migrant Krobo farmers, in portions of Amanase-
Whanabenya settled by Shai migrant farmers, and in other areas settled by
these people. Explanation lies in the *huza* system. The *huza* land is shared
into individually owned longitudinal strips called *zugba*. Homes, or dwelling
units, are constructed at the base of each *zugba*, giving rise to a linear settlement pattern that contrasts with the nucleated pattern typical of Akan areas.
From the home area, farming proceeds in the same general direction along
the *zugba*, uninhibited by other dwelling units, unlike what frequently happens in built-up nucleated settlements. Gardens, commonly of the
agroforestry type, are developed within a few hundred metres from the *zugba*
home. Beyond lie other land-use types including non-home garden agroforests (Field, 1943; Gyasi, 1976).

The home garden agroforestry type contains virtually all the varieties of
crops found in the non-home garden type, and more. Although cocoa has generally diminished in importance, it still shows higher concentrations in the
home garden agroforests. Other crops more common in home garden agroforests include:

○ peppers (*Capsicum annuum*)
○ condiments and certain types of leafy vegetables in regular demand for the
  kitchen
○ plantains and bananas
○ fruit trees such as mango, avocado, citrus, coconut and *Chrysophyllum
  albidum* (called *adesaa* by Twi-speaking Akan people and *alatsa* by
  Adangbe-speaking Krobo and Shai people)
○ yams and taro that thrive in the shady and humid conditions created by the
  tree canopy.

Agrobiodiversity can be high even in the agroforests of tenant farmers, who are
often assumed to be less motivated to conserve biophysical resources. Species
found in the home garden agroforestry unit of a migrant settler tenant farmer
at Otwetiri in Gyamfiase-Adenya are shown in Table 24.2 (Gyasi, 1999).

Table 24.2   Plants found in an agroforestry home garden of a migrant settler tenant farmer during a PLEC survey in 1998/99 at Gyamfiase-Adenya, Ghana

Crop Twi name	English or Scientific name	Other plants (sapling or tree) Twi name	Scientific name
Bankye	Cassava	Odwen	*Baphia nitida*
Aburow	Maize	Akakapenpen	*Rauvolfia vomitoria*
Afasew	Water yam (*Dioscorea alata*)	Onyankyren	*Ficus exasperata*
Amankani	Taro (*Xanthosoma*)	Osena/Yooye	*Dialium guineense*
Brode	Plantain	Ankye	*Blighia sapida*
Brofere	Papaya	Abrewa aninsu	*Hoslundia opposita*
Abrobe	Pineapple	Osisriw	*Newbouldia laevis*
Mmofra ntorewa	*Solanum torvum*		*Bryophyllum pinnatum*
Mako	Peppers		

Source: Gyasi (1999).

In most *huza* areas crop yields are sustained, at least partially, by the year-round addition of household refuse. In valleys where many of the homesteads are located, an additional source of plant nutrients is from soils washed down from uphill.

Yam diversity is of particular importance in Ghana, as the region is a centre of diversity for this tuber. Yams continue to feature significantly in the diet and commerce of the forest-savanna zone, especially in migrant Krobo areas. Studies have shown that six major types are commonly grown at the study sites (Table 24.3). The number of varieties identified by their local names found in each site are shown in Table 24.4. At Sekesua-Osonson (a migrant Krobo area) farmers named 54 varieties, and at Gyamfiase-Adenya (occupied by a mix of Akuapem people and more recent migrant Ayigbe and Ewe settler-tenant farmers from Togo and Ghana's Volta Region) they named 53. Thirty-six varieties were named in Amanase-Whanabenya, which is similarly settled by a mix of Akuapem, Shai, Ayigbe and Ewe migrant farmers.

Table 24.3   Major types of yam (*Dioscorea* spp.) grown in the study sites of Sekesua-Osonson, Gyamfiase-Adenya and Amanase-Whanabenya, Ghana

Common name	Botanical name	Local name
White yam	*D. rotundata*	Ode (Akan), Hier (Adangbe)
	D. rotundata (wild)	Kookoo ase bayere (Akan), baale (Adangbe)
Water yam	*D. alata*	Afasew (Akan), Alamoa (Adangbe)
Yellow yam	*D. cayenensis*	Nkani (Akan), Kani (Adangbe)
Bitter yam	*D. dumetorum*	Nkamfoo (Akan)
Chinese yam	*D. esculenta*	Potato (Colloquial), Aduanan (Akan)
Aerial yam	*D. bulbifera*	Tshomkani (Adangbe) Osowaba (Akan)

Source: Adapted from a report by Blay (2001) based on PLEC field studies.

Table 24.4   Number of varieties of each yam (*Dioscorea*) species in the Gyamfiase-Adenya, Sekesua-Osonson and Amanase-Whanabenya study sites, Ghana

Species	Number of varieties		
	Amanase-Whanabenya	Gyamfiase-Adenya	Sekesua-Osonson
White yam (*D. rotundata*)	18	28	26
Water yam (*D. alata*)	11	14	15
Yellow yam (*D. cayenensis*)	3	4	5
Bitter yam (*D. dumetorum*)	2	4	6
Chinese yam (*D. esculenta*)	1	1	1
Aerial yam (*D. bulbifera*)	1	1	1
	36	52	54

Source: Adapted from a report by Blay (2001) based on PLEC field studies.

Yam management, especially by Krobo farmers, is characterized by practices that conserve soil and agrodiversity. The method of minimal tillage avoids construction of mounds by sowing directly into a hole drilled in the soil with little disturbance before sowing. The yam hole is then dressed using decomposing leaves, cocoa husk and other biomass to enrich the soil. Live-staking makes use of various plants, particularly *Newbouldia laevis*, called *nyabatso* by Adangbe-speaking people, and *osisriw* by Twi-speaking Akuapem people. Growing a diversity of yam varieties staggers the harvest. Smaller yam tubers are then stored after the harvest *in situ* in the soil to serve as seed stock.

Non-agroforestry home gardening

A third major traditional way of conserving agricultural biodiversity is in home gardens that exclude trees. Common plants in the home garden include vegetables, legumes, condiments that are in frequent demand for the kitchen, *Musa* species especially plantains, cassava, and herbaceous medicinal plants that often must be fetched at short notice to meet emergency medical situations. Although non-agroforestry home gardens still persist, close spacing of housing associated with population growth and a lack of settlement planning is increasingly crowding them out in the nucleated villages. A 1998/99 survey showed the number of all types of home garden per compound house to be significantly greater in linear settlements than in nucleated ones (Table 24.5).

Threats to traditional farming systems

While traditional systems that secure both biodiversity and food production endure, they appear threatened. In the 2000 PLEC survey, over 30 per cent of a sample of farmers in the study sites perceived a threat to diversity of crops, and over 80 per cent to diversity of natural trees including those associated with crop-farming. Foremost among the crops under threat was taro, followed by plantains and yams. The perceived threat was most acute in Gyamfiase-Adenya and Amanase-Whanabenya.

Table 24.5   Concentration of home gardens per compound house in non-nucleated linear settlements relative to nucleated settlements in PLEC study sites

Study site	Settlement name	Home garden concentration (Number of home gardens per compound house)	
		Non-nucleated	Nucleated
Amanase Whanabenya	Whanabenya-Nyamebekyere	1.55	
	Abenabo	0.93	
	Obongo	1.15	
	Aboabo		0.75
	Amanase		0.80
	Aye-Kokooso		1.22
Sekesua-Osonson	Osonson Korlenya	2.20	
	Prekumasi	1.00	
	Siblinor	1.30	
	Sekesua		1.15
Gyamfiase-Adenya	Otwetiri		0.46
	Kokormu		0.39
	Yensiso		0.70
Average		1.36	0.78

Source: A 1998/99 PLEC survey

Expansion of modern agricultural practices and their local adaptations threaten traditional systems. Modern agricultural enterprises comprised nearly 28.6 per cent of the total sample of farms surveyed by PLEC at these three sites and an additional one in the savanna-forest zone in 1995. Modern enterprises comprised:

○ 12.8 per cent small-scale monoculture
○ 11.6 per cent ranching
○ 2.2 per cent large-scale plantations
○ 1.0 per cent intensive battery-style poultry farming
○ 0.2 per cent market gardening
○ 0.8 per cent other.

Land shortage associated with population pressure partly underlies the threat. A desire to maximize economic returns from limited land encourages monoculture, especially by tenants, because of exacting rents or unfavourable share-cropping arrangements (Gyasi, 1976, 1994, 1998).

Farmers stress declining crop productivity from limited land. Their ability to counter the problem is constrained by poverty and limited access to financial resources, especially credit. A lack of awareness of the intrinsic and potential commercial value of agricultural and biological diversity is also a factor. At present there is weak demand for many of the diverse landraces, especially of the traditional crops that farmers produce. There is growing preference for imported exotic food items, notably wheat and American long-grain rice. The changing preference is particularly conspicuous among the young people, who rarely value the survival of traditional crop diversity. Another factor is the hasty introduction of modern farm management practices without an adequate appreciation or understanding of the locally adapted traditional practices that the new practices are intended to replace.

Countering threats

An explicit policy by the government of Ghana to encourage biodiversity conservation in the country is still in draft stage (Ministry of Environment, Science and Technology, 1998). However, the Medium Term Agricultural Development Programme implicitly addresses the biodiversity issue by seeking accelerated, ecologically harmonious sustainable agricultural modernization and production. It emphasizes organic farming not only because it is known to farmers, but also because it is 'cheap, benefits soil structure and soil moisture-holding capacity, reduces the need for chemical fertilizers and helps to increase the efficiency with which these are used by the plant' (Republic of Ghana, 1990: 64).

Recognition of a need to protect biodiversity was demonstrated by the establishment by government of a National Plant Genetic Resources Centre under the Council for Scientific and Industrial Research. This aims to conserve plant genetic resources through both *ex-situ* and *in-situ* conservation methods in collaboration with farmers. Non-governmental organizations, including PLEC, are also involved in biodiversity conservation. PLEC seeks optimal paths of conservation of biodiversity within agriculture. Work has focused not only on development of demonstration sites at Gyamfiase-Adenya, Sekesua-Osonson and Amanase-Whanabenya in the southern sector of the forest-savanna ecozone, but also at Jachie in central Ghana, and Bongnayili-Dugu-Song in the northern sector. Subsidiary sites include Bawku-Manga in the north (Figure 24.1).

Over the past eight years, particularly since 1998, studies have been carried out, in association with farmers, on agroenvironmental changes, smallholder land-use practices, and strategies for countering environmental degradation. Inventories of crops and plants, including rare varieties of yam, and traditional methods of sustaining biodiversity were compiled. There has been increased appreciation by farmers of stone lining as a technique of minimizing soil erosion along slopes, and of the efficacy of *oprowka*, a traditional system of soil and biota conservation by not burning cleared vegetation but leaving it in place to decompose for mulch. There has been an increase in popular awareness of trees that combine effectively with food crops following an experiment based on traditional agroforestry practice (Owusu-Benoah and Enu-Kwesi, 2000). Demonstration activities have resulted in the conservation of

○ trees alongside crops in farms
○ forests adjoining farms
○ threatened local varieties of rice and domestic fowl
○ medicinal plants in arboreta
○ plants in school gardens and in farms owned privately or on a group basis.

The activities were facilitated by farmer visits and exchange of germplasm, and by the strategy of using expert farmers for demonstration and extension within and between demonstration sites.

Income-generating activities have been initiated, including some aimed at adding value to agrobiodiversity. Farmers have used forests conserved nearby for snail farming and bee-keeping for honey and wax. They have raised seedlings on a commercial basis as individuals, families, and through the farmers' associations. A piggery was founded partly using residues from farming and food preparation. An association of female farmers began processing cassava into flour for bread and pastries, and other women began cotton spinning and weaving and dry-season production of local vegetables.

Conclusion

Biodiversity within agriculture still remains high in Ghana because of its dynamic character. However, pressures of production and forces of modernization threaten this resource, which is crucially important for food security. One possible way of stemming the threat is to strengthen the national capacity for conservation through popular programmes and training, and by using a participatory approach modelled on the successful PLEC initiative.

Part IV. The way forward

25 The message from PLEC[1]

HAROLD BROOKFIELD

THIS FINAL SHORT chapter sets out to distil the message of PLEC to all interested readers. I need to recapitulate that not only can biodiversity be cultivated, but a great many farmers in the world in fact do this. This is evident from the whole book, but rather than develop it further, this chapter is instead about what the method and experience of PLEC can teach in the areas of rural development and resource conservation. PLEC is not just another in the modern wave of 'development with conservation' projects. In what is, methodologically, the most important chapter in this book, Pinedo-Vasquez et al. (Chapter 10), laid stress on the integration of development and conservation goals within one set of activities, combining rather than trading off different objectives.

The methodological lessons of PLEC

In an article on the field of environment–development work, Kirkby et al. (2001) classify all those development approaches that emphasize community-based participation, are committed to local diversities, and involve analysis and action centred on the grass roots, as either 'neo-populist' or 'eco-populist'. PLEC does all these things, but few of its members would regard themselves as 'neo-populists' or 'eco-populists', and the reasons are of some importance. The project's genesis is different from that of most others.[2]

PLEC has been a project built up by scientists, most of them natural scientists, working in the unfamiliar milieu of rural development and biodiversity conservation. In the process, they have had to adapt themselves to the practice of working with farmers, and listening to them, rather than merely observing what they do and talking to them. They have come into this work from scientific research, and for many of them being willing to learn from farmers has required a major shift in their way of thinking. By no means all have succeeded, but for those who have, the shift has had many rewards. The scientists have often made close friends in the villages and have gained co-operation in ways not possible by standard development interventionism. Their own research opportunities have been widened as a result. Some scientists have been very successful in adopting and benefiting from new ways of working, and those who have made this shift are well represented among the authors in the present collection. Success has called for open-minded people, who have allowed the field experience and the farmers to teach them, while offering their own scientific expertise and skills in return.

There has certainly been an evolution of method as increasing emphasis has been placed on the role of expert farmers, even a revolution in cases such as in Tanzania in 1998 (Chapters 12 and 14) and among the group working at Baihualing in China in 1999 (Chapter 21). Amazonia adopted what is now the PLEC approach from the outset, although they have since refined it (Chapters 15, 16 and 17). A progressive and entirely spontaneous evolution is exemplified in West Africa (Chapters 13, 23 and 24). This has happened also in areas not represented here. Success has been more mixed at some sites, not always due to the reluctance or inability of scientists to change their ways. There are regions of the world in which patterns of interaction among farming families do not allow for any variant of PLEC's farmer-to-farmer training approach. In those instances our colleagues used or developed other methods.

One common trend, however, has been a shift away from the standardized 'participatory' tools common in development circles since the late 1980s, growing out of farming-systems research and now particularly characteristic of what is called 'neo-populist' work. Many of our members did start off with such devices as rapid rural appraisal, participatory rural appraisal, participatory planning and action, and focus group discussions.[3] They moved on from these limited ways of seeking consensus after they began to identify the expert farmers, and to realize how much could be achieved by encouraging selected individuals. These experts become farmer-instructors, and the knowledge gained from them is 'used by farmers to enhance their production, by project directors to advance their goals of implementing development with biodiversity conservation, and by PLEC scientists to further develop our methods' (Pinedo-Vasquez et al., Chapter 10, p. 107).

Core elements of the PLEC message

The PLEC message has thus acquired a distinctive nature, and it includes the following elements:

○ The core of successful conservation-with-development is the identification of those farmers and other resource managers who do the job best, and in building extension among other farmers on these people and their practices.

○ Such extension needs to have short-term tangible benefits to the farmers and others, as well as long-term conservation benefits. Where biodiversity conservation is a long-term goal, activities that create value out of biodiversity need to be isolated and assisted. Two particularly striking illustrations, from central Ghana and Guinée, are presented in Chapters 13 and 23.

o The identification of the best resource managers and the right activities to assist calls for scientific skills as well as ability to listen and observe. It is not easy, and it takes time.

o The practices of the best are not only diverse but also dynamic. Ability to innovate, adapt and change has been encountered in all areas, and is a major point of strength both for development and conservation.

o The concentration of work in a limited number of demonstration sites has had important advantages. PLEC has now worked in some of them for several years. In-depth observation, collaboration and monitoring are facilitated, and work can continue to evolve around the knowledge, experience and inventiveness of the farmers themselves. Exchange-visiting by farmers between the demonstration sites has accelerated progress in all. The more successful demonstration sites have become centres for the diffusion of ideas, and of successful innovations, into widening areas. The case study chapters present clear illustrations of these possibilities from Amazonia, West Africa and Tanzania.

o The PLEC approach is a method, not a formula. What is done at these sites varies substantially between areas even within the same country, being sensitive to local conditions of environment, society and external conditions. Demonstration activities are continuously evolving. In general, a capacity for farmer-to-farmer instruction is a consequence. The necessary role of the scientists does not fade away, but it becomes supplementary. Farmers see and seize new opportunities, and these opportunities include benefits from the conservation and even enrichment of biodiversity.

o In human resource terms, PLEC is inclusive rather than exclusive, and regular agricultural extension staff, and members of other projects, have increasingly participated in farmer-led activities. Many have found the experience stimulating and challenging. A greater confidence in ability to manage diverse resources is thus spread from the village to the extension staff, and to other areas.

Biodiversity management at landscape level

Although some of the case studies presented here have been at the level of particular fields or plots of forest, the majority have been at landscape level, as defined in Chapter 3. It has been necessary to bring together the managers of a range of agroecosystems, and their associated managed fallow and wild areas, in order to devise strategies for the productive management of biodiversity. Sometimes this could be done through existing local institutions, but a useful device developed in a number of areas has been the 'farmers' association', formed by the farmers themselves to manage their work with PLEC.

This sort of specific-purpose farmer cooperative is, in effect, an adaptation to local conditions of the associations of farmers formed to resolve landscape-wide conservation and management problems in some more developed agricultural countries. A particular case in point is the several thousand 'landcare' groups formed in Australia and New Zealand since the 1980s, and more recently also formed in South Africa. They mostly consist of neighbouring farmers working together to solve common problems such as weeds, erosion, soil salinity and feral animals. They often map their own soil resources and problem sites, and develop cooperative plans for management. Maintenance and encouragement of biodiversity often form a part of their activities, though it is rarely central.

These groups, like the PLEC farmers' associations, are what are termed community-based organizations (CBOs), even though there may often be no 'community' other than the one formed on the spot. In wealthier countries such organizations receive considerable scientific support, and some financial support also, from resources in the control of the state and its agencies. This is not possible in poorer countries. PLEC scientists have filled the gap, bringing in some help from government agencies and NGOs that would not otherwise be attracted to give support. But the principal sources of income have been from the local enterprises of the farmer's associations themselves. PLEC has provided minimal financial support.

This sort of local organization can also take up and help popularize local conservation initiatives, like the reservation of small forest patches, the replanting of degraded watershed protection woodland, and – in a specific Tanzanian case described in Chapter 12 – a woodlot in which medicinal, fuelwood, nitrogen fixing and food- and fodder-yielding species were conserved. Local conservation initiatives are seldom recognized for what they are by outside observers, and may even be resented by neighbours. Local empowerment is the most effective means of supporting them. However, the scientists are also needed.

A final recommendation

The homes of all but a few Cluster scientists are in the universities and research institutions of the developing countries. One conclusion to be drawn from this book is that these institutions clearly contain a large resource of skilled and adaptable people, capable of developing a practical project that integrates conservation with development. This constituency has been drawn on little in the past by projects emanating from their colleagues in developed countries. PLEC has abundantly demonstrated the value of this pool of scientists, and their ability to find the best farmers and resource managers, and endow them with responsibility and the confidence that comes with it.

Spread out to the local level, the financial resources needed to continue PLEC-like work drawing on this willing expertise are quite modest.

Collectively, PLEC has been a large project, but most of its parts are capable of continuing by themselves, or in regional consortia. To the wider global field of organizations specializing in development and conservation, and especially those specializing in the funding of such work, we recommend this resource as a great pool of skills and abilities on which it is possible, given guidance rather than direction, to draw. A more sensitive approach to the problems of the world's small farmers might well emerge, and quite quickly, by so doing.

Contributors

Dr Kojo S. Amanor is Senior Research Fellow in the Institute of African Studies at the University of Ghana, Legon, Accra, Ghana

M. Alhassane Baldé is Head of the Soil Conservation Department at the Centre de Recherche Agronomique, Bareng, Pita, BP 1523 Boulevard de Commerce, Conakry, République de Guinée

M. Abdoul Karim Barry is Lecturer and Manager of the Central Laboratory, at the Centre d'Études et de Recherche en Environnement, Université de Conakry, BP 3817, Conakry, République de Guinée

Professor Ibrahima Boiro is Leader of the PLEC Guinée Cluster and Director of the Centre d'Études et de Rechèrce en Environnement, Université de Conakry, BP 3817, Conakry, République de Guinée

M. Omar Barry is a collaborating expert farmer at Dar ès Salam, Bantignel, Préfecture de Pita, République de Guinée

Professor Harold Brookfield has been Principal Scientific Coordinator of PLEC from 1992 to 2002, and is a Visiting Fellow in at the Department of Anthropology, Research School of Pacific and Asian Studies (RSPAS), Australian National University, Canberra, ACT 0200, Australia

Ms Muriel Brookfield is Editor of *PLEC News and Views*, and a Departmental Visitor at the Department of Anthropology, Research School of Pacific and Asian Studies (RSPAS), Australian National University, Canberra, ACT 0200, Australia

Mr Chen Aiguo, Head of the Xishuangbanna Group of PLEC China, is Associate Professor in Agroforestry at the Xishuangbanna Tropical Botanical Garden (Kunming Office), The Chinese Academy of Sciences, No 50 Xuefu Road, Kunming, Yunnan 650223, China

Mr Kevin Coffey, a member of PLEC's Scientific and Technical Advisory Team, is Research Assistant in the Institute of Economic Botany, New York Botanical Garden, Bronx, New York, NY 10458–5126, USA

Mr Cui Jingyun is Research Associate in Botany at the Xishuangbanna Tropical Botanical Garden, The Chinese Academy of Sciences, Menglun, Mengla County, Xishuangbanna, Yunnan 666303, China

Mr Dao Zhiling, Head of the Gaoligongshan Group of the PLEC China Cluster, is Associate Professor in the Kunming Institute of Botany, The Chinese Academy of Sciences, Heilongtan, Kunming, Yunnan 650204, China

Dr Amirou Diallo is Head of the Department of Biodiversity and Environmental Impact Assessment, at the Centre d'Études et de Rechèrce en Environnement, Université de Conakry, BP 3817, Conakry, République de Guinée

Mr Du Xiaohong is Director of the Baoshan Forestry Aviation Station, SW Forestry Aviation General, State Forestry Bureau, Shenyang Donglu, Baoshan, Yunnan 678000, China

Associate Professor Lewis Enu-Kwesi is Deputy Leader of the Ghana Cluster of PLEC, and a member of PLEC's Scientific and Technical Advisory Team. He is Head of the Department of Botany, University of Ghana, PO Box 55, Legon, Accra, Ghana

M. Sékou Fofana, formerly of the Centre d'Études et de Rechèrce en Environnement, Université de Conakry, is presently Technical Coordinator of the Projet Elargi de Gestion des Ressources Naturelles, Winrock International, BP 6575, Conakry, République de Guinée

Mr Fu Yongneng is Assistant Research Officer in Agroforestry at the Xishuangbanna Tropical Botanical Garden, The Chinese Academy of Sciences, Menglun, Mengla County, Xishuangbanna, Yunnan 666303, China

Ms Guan Yuqin, formerly a graduate student of the Kunming Institute of Botany, Chinese Academy of Sciences, is now Deputy Head of the Vocational School of Guanzhou City, China

Professor Guo Huijun is Leader of the PLEC China Cluster and Vice-Director of the Xishuangbanna Tropical Botanic Garden, The Chinese Academy of Sciences, Menglun, Mengla County, Xishuangbanna, Yunnan 666303, China

Professor Edwin A. Gyasi is Coordinating Leader of PLEC in West Africa and Professor of Geography at the Department of Geography & Resource Development, University of Ghana, PO Box 59, Legon, Accra, Ghana

Mr Fidelis B.S. Kaihura is Cluster Leader of PLEC Tanzania, and is a Senior Agricultural Research Officer at the Agricultural Research and Training Institute Ukiriguru, PO Box 1433, Mwanza, Tanzania

M. Mohamad Aliou Kane is a collaborating expert farmer at Missidè Héïré, Bantignel, Préfecture de Pita, République de Guinée

Mr Edward Kemikimba is a Senior Agricultural Field Officer at the Agricultural Research and Development Institute Ukiriguru, PO Box 1433, Mwanza, Tanzania

Mr Chawalit Korsamphan is a researcher at the Highland Agricultural Research and Development Centre, Faculty of Agriculture, Chiang Mai University, Chiang Mai 50200, Thailand

Mr Liang Luohui is Managing Coordinator of PLEC in the Academic Division, United Nations University, 53–70 Jingumae 5-chome, Shibuya-ku, Tokyo 150–8925, Japan

Mr Li Yingguang is Office Manager at the Baoshan Nature Reserve Bureau, Baoshan County, Lujiang Township, Baoshan County, Yunnan 678028, China

Dr David McGrath, Deputy Leader of the PLEC Cluster in Brazilian Amazonia, is Vice-coordinator of the doctoral programme at the Nucleo de Altos Estudos Amazônicos (NAEA), Campus Profissional Guama, Universidade Federal do Pará, CEP 66075–900, Belém, Pará, Brazil

Ms Fanta Mara is Head of the Department of Geographical Information Systems and Remote Sensing, at the Centre d'Études et de Recherche en Environnement, Université de Conakry, BP 3817, Conakry, République de Guinée

Mr Paulo Ndondi is a Senior Land Use Planning Officer at the Department of Research and Development Headquarters at the Ministry of Agriculture and Food Security, PO Box 2066, Dar es Salaam, Tanzania

Dr William Oduro is Group Leader of the Central Ghana PLEC team. He is Senior Lecturer at the Institute of Renewable Natural Resources, Kwame Nkrumah University of Science and Technology, University Post Office, Kumasi, Ghana

Dr Christine Padoch has been Associate Scientific Coordinator of PLEC from 1993 to 2002. She is Matthew Galbraith Perry Curator of Economic Botany at the Institute of Economic Botany, New York Botanical Garden, Bronx, New York, NY 10458–5126, USA

Ms Helen Parsons is Research Assistant to the PLEC Project, Department of Anthropology, Research School of Pacific and Asian Studies (RSPAS), Australian National University, Canberra ACT 0200, Australia

Dr Mario Pinedo-Panduro, field coordinator of PLEC in Peru, is Research Scientist in the Terrestrial Ecosystem Programme at the Instituto de Investigaciones de la Amazonia Peruana (IIAP), Casilla Postal 471, Iquitos, Peru

Dr Miguel Pinedo-Vasquez, Convenor of PLEC's Scientific and Technical Advisory Group, is Associate Research Scientist in the Center for Environmental Research and Conservation, Columbia University, MC 5557, 1200 Amsterdam Avenue, New York, NY 10027–6902, USA

Mr Fernando Rabelo is a researcher at the Instituto de Pesquisa Ambientales da Amazônia (IPAM), Avenida Nazaré 669, Barrio de Nazaré Cep–66035–170, Belém, Pará, Brazil

Professor Benjavan Rerkasem is a Senior Scholar of the Thailand Research Fund, and Professor of Plant Nutrition in the Faculty of Agriculture, Chiang Mai University, Chiang Mai 50202, Thailand

Dr Kanok Rerkasem, Head of the PLEC-Thailand Cluster, is a research scientist at the Multiple Cropping Centre, Faculty of Agriculture, Chiang Mai University, Chiang Mai 50200, Thailand

Professor Michael Stocking has been Associate Scientific Coordinator of PLEC from 1993 to 2002. He is Professor of Natural Resources in and, from 2003, Dean of the School of Development Studies, University of East Anglia, Norwich NR4 7TJ, United Kingdom

Mr Charal Thong-Ngarn is an officer in the Highland Agriculture and Social Development Promotion Office, in the Department of Public Welfare, Chiang Mai 50200, Thailand

Dra Tereza Ximenes-Ponte, Leader of the PLEC Brazilian Amazonia Cluster, is Professor of Interdisciplinary Methodology and coordinator of the graduate programme at the Nucleo de Altos Estudos Amazônicos, Campus Profissional Guama, Universidade Federal do Pará, CEP 66075–900 Belém, Pará, Brazil.

Mr Narit Yimyam is a researcher in the Highland Agricultural Research and Development Centre, Faculty of Agriculture, Chiang Mai University, Chiang Mai 50200, Thailand

Dr Daniel J. Zarin is Associate Professor of Tropical Forestry at the School of Forest Resources and Conservation, and is Executive Director of the Forest Management Trust, University of Florida, PO Box 110760 Gainesville, FL 32611, USA

Notes

Chapter 2

1 A succinct definition of a 'landrace' is that provided by Sauer (1993: 275): 'a sexually reproducing crop variety developed under natural and folk selection'.
2 Respectively, these are entitled 'Strengthening the Scientific Basis of *in situ* Conservation of Agricultural Biodiversity On-farm', and 'The Community Biodiversity Development and Conservation Programme'.

Chapter 3

1 Another example of such a 'precision' system, from outside PLEC, is found in work on indigenous soil conservation in a part of Burkina Faso by Mazzucato and Niemeijer (2000). Setting out to find physical soil and water conservation works, they found little reliance on these in achieving successful management. Field management practices integral to the whole farming system seemed to have been primarily responsible for maintenance of the condition of the land. The farmers they studied understood very well that soil is created by erosion, sedimentation and the incorporation of organic matter. However, they had limited quantities of the inputs necessary to manage these processes, including their own time. They therefore had to make great use of skills and learning. They applied physical and biological management adaptively in relation to site conditions and the conditions of the season. In the process, soil quality was improved over time.
2 In Asia, it became part of a 'sloping agricultural land technology' (SALT). The International Board for Soil Research and Management (IBSRAM), as it was called until 2001, also advocated alley-cropping as a principal strategy in sloping land management.
3 In the Burkina Faso example mentioned above, farmers invested both time and goods in maintaining and widening these networks, and used them in borrowing land so as not to have to overuse their own (Mazzucato and Niemeijer, 2000).
4 PLEC has had little historical dimension. Documentary material is very limited in most Cluster areas. However, the history of land use was pursued in Amazonia, China and West Africa, and Chapters 7, 15, 17, 23 and 24 in this book provide historical information.

Chapter 4

1 This chapter is a substantially revised version of a presentation given to the meeting of the International Geographical Union's Commission on Land Degradation in Perth, Western Australia in 1999. An earlier version appeared in Stocking (2000)
2 Notwithstanding the title of the Convention, the term 'desertification' is increasingly being avoided. Pagiola (1999) cites two reasons: the vagueness and imprecision of its definition, and the controversy over whether a process which has its major emotional impact in images of advancing sand dunes should be described as a real phenomenon. The GEF Operational Programmes also studiously avoid the word.

Chapter 5

1 In a pioneer study of an African system, still of value despite its age, de Schlippe (1956: 117–18) speculated as follows on how such Field Types arose: 'Theoretically, one could think of thousands of different ways in which the great number of crops and varieties and the astonishing mosaic of soil-vegetation types could be combined into field types. In practice one discovers, however, that field types are few and that it is always the same field types that are repeated by all members of the group'.

2 Methods of evaluating the resource endowment of farm households, and of differentiating farm households by their resource endowment, have a large literature, surrounded by some controversy. The topic is one for separate discussion. PLEC advocates the Amazonian model in which selection focuses on expert farmers who farm and conserve best, are the most imaginatively experimental and innovative, and who can teach others. Although all PLEC's farmers are small farmers, not all are equal. Where there are large differences between richer and poorer farmers, some ranking is desirable.

3 An example in which data were collected and analysed by members of the East Africa Cluster is set out in Tengberg et al. (1998).

Chapter 6

1 The final paragraph is reproduced from Zarin (1995: 19–20).

Chapter 8

1 This chapter was written on behalf of the Scientific and Technical Advisory Team (STAT) of PLEC which, in addition to Kevin Coffey, includes Miguel Pinedo-Vasquez, Edwin Gyasi and Lewis Enu-Kwesi.

2 The data used here, and the results of the index calculations, are discussed in Chapter 15.

3 While the natural log is most commonly used in this calculation, any base can be used.

4 The Shannon Index weighs both species richness and evenness. There are also indices that measure only evenness. See Magurran (1988)

5 Measures can be estimated from a rank abundance graph when they are compared to a set of distributions, such as log normal. See Magurran (1988).

6 An example of the use of the Jaccard coefficient and cluster analysis is in Chapter 20.

7 In average linkage clustering, new similarity measurements are recalculated for each new cluster by averaging the similarity measures of the samples within the cluster.

8 Eigenvalues are measures of the relative importance of each component. The higher the eigenvalue, the more representative the principal component is of the variation in the original data. Eigenvalues quantify the principal component's likeness to the data set.

Chapter 9

1 For a more complete description of this case, see Padoch et al. (1998).

Chapter 10

1 The authors are members of the project's Scientific and Technical Advisory Team. During 1999 and 2000, Pinedo-Vasquez and Gyasi, accompanied in part by Coffey, visited Clusters in both West and East Africa, China, Thailand and Papua New Guinea, and all met in Brazil during a general project meeting in 2000. Most of the illustrative material is drawn from Amazonia (Brazil and Peru) and Ghana, reflecting the principal experience of the authors. The clear implications in the text that some of PLEC's Clusters have not attained the high standards set in this chapter are correct, but the authors have refrained from identifying which they are.

Chapter 11

1 Some sacred trees protected include odii (*Okoubaka aubrevillei*) and ahomakyem (*Spiropetalum heterophyllum*). Before farmers can collect any part of these trees for medicinal or other use they must first perform an elaborate series of rituals, which include the pouring of libation and the sacrifice of eggs.

Chapter 14

1 Soil properties were measured by excavating 10×50 cm^2 mini-pits. Soils were described in terms of mini-pit characteristics and named in local and scientific names. Other descriptive parameters included: topsoil depth, surface (0–20 cm) and subsurface (30–50 cm) soil properties of colour, texture, structure, consistency, pore size and distribution, and root size and abundance. For each mini-pit soil samples were taken from surface and subsurface horizons for laboratory analysis. Results are reported elsewhere.

2 *Mbugas* are low-lying parts of the landscape, seasonally waterlogged, consisting of peats (Histosols or organic material) and Vertisols (active clays). In many catchments, they are the principal source of dry-season grazing, or where population density is greater, the main place where vegetables are grown because of the good moisture and high intrinsic fertility. There are quite a number of types of *mbuga* depending on characteristics such as the degree of wetness, heaviness of the clay, degree of salinity, and month in which it dries sufficiently to plant vegetables.

Chapter 15

1 Calculations of the indices used are set out in Chapter 8. In this case we estimated β-diversity by dividing the number of species found in the sites by the number of species in each landholding.

Chapter 16

1 The authors are principal members of the PLEC Cluster in Peru, but the work they describe here began before PLEC. It continues in association with PLEC, which undertakes wider activities in the same Muyuy area. This chapter gives a valuable demonstration of the wider applicability of what has become the PLEC method of working with farmers.

Chapter 17

1 PLEC began work in Amazonian Brazil in 1992, and work in Amapá, already begun under another project, was then incorporated into PLEC. The task reported here overlaps the period of the two projects.
2 For more complete discussions of the formation of the smallholder timber industry see Pinedo-Vasquez et al. (2001) and Sears et al. (2000).

Chapter 18

1 Translated by Liang Luohui, Managing Coordinator of PLEC.

Chapter 19

1 This was a student paper when it was written in the mid-1990s. Even though there have been some changes in the situation at Baka, and in the market for *Amomum villosum*, since that time, the paper is reprinted without substantial change.
2 55 000 mu is approximately 3700 hectares; 1 ha is equivalent to 15 mu.

Chapter 20

1 We thank San Long, and Mi Ba for helping us in the field work. Liang Luohui has given invaluable assistance with finalization of the English version of this paper. Kevin Coffey advised on the use and interpretation of the indices employed.

Chapter 21

1 One hectare is equivalent to 15 mu.

Chapter 22

1 The authors wish to acknowledge support for part of the work reported in this chapter from UNU-PLEC and Thailand Research Fund
2 Karen, Hmong, Lahu, Lisu are some of the most common ethnic minority groups living in the mountainous areas of mainland Southeast Asia.
3 Although some groups, including the Karen, have been in Thailand for several hundred years, others are migrants within the twentieth century.
4 Use of these edges between fields is a particularly distinctive feature of the Pah Poo Chom system. A few are substantial, with natural as well as planted trees. The crops and other products obtained from them are mainly used for self-provisioning, and in this respect they constitute an important diversification of what is otherwise largely a commercial system of land use.
5 At the time of this research US$1 = Bt43.
6 Research into Pada currently being undertaken includes its biology of seed production and dormancy, its contribution to productivity of upland rice in the cropping system, its role in nutrient cycling and relationship with other key fallow species. The role of mycorrhiza and nitrogen-fixing endophytes in the nutrition of Pada is also being investigated. This work is supported by UNU-PLEC and Thailand Research Fund.

Chapter 23

1 Use of the name *runde* for villages inhabited by former slaves was officially forbidden after the land reform made these the property of their occupants, making all communities equal in the eyes of the law. In practice, however, many older rights of the big landowning families are still recognized. Although many former slaves have been able to buy their family concessions (*sunturés*), others still have to pay 10 per cent of the grain crop to their former masters.
2 In each sector there is a considerable number of other named places, referred to in our reports and in other documents. For simplification here we use only the sector names.
3 The *farilla* is paid by all farmers, but those who have no landlords pay it to the mosque.

Chapter 24

1 Taro (*Colocasia* spp.) is generally known in Africa as the cocoyam, an ancient name. Nonetheless, the more international 'taro' is used here.

Chapter 25

1 Although this chapter is written by the principal editor, there have been substantial inputs at draft stage from both Christine Padoch and Michael Stocking. At their suggestion, however, only one author's name appears.
2 In a broader review of the state of the development literature, and that of its critics, as a whole, Blaikie (2000: 1045) describes as not only 'neo-populist' but also 'post-modern', writing that, inter alia, it rejects modernization, respects diversity and local agendas, and encourages authentic local action so that people can speak and act for themselves. However, no one in PLEC has so shifted the focus of their work away from the farmers, and fellow professional scientists, as even to consider identifying themselves among the academic movement of 'post-modernism'. Most might even object, quite strenuously, to having their work so classified.
3 'Participatory' methods as understood in development usually require participation in someone's agenda, most often that of the project personnel. Many recent 'integrated conservation and development projects (ICDPs)' also seek to derive a way of integrating conservation and development within one set of activities, by combining local and global objectives, not trading them off. Unfortunately, they often have a strong ideology from formal scientific ecology, and have been bedevilled in practice by conflicting objectives. Their agendas have been beset by intractable differences between what local people perceive that they need and what scientists are prepared to give.

References

Aarnink, W., Bunning, S., Collette, L. and Mulvany, P. (1999) *Sustaining Agricultural Biodiversity and Agro-ecosystem Functions*. FAO, Rome.

Abdulai, A.S., Gaysi, E.A., Kufogbe, S.K., with Adraki, P.K., Asante, F., Asumah, M.A., Gandaa, B.Z., Ofori, B.D. and Sumani, A.S. (1999) Mapping of settlements in an evolving PLEC demonstration site in northern Ghana: an example in collaborative and participatory work. *PLEC News and Views* 14: 19–24.

African Development Bank (1995) *Selected Statistics on Regional Member Countries*. African Development Bank, Abidjan, Côte d'Ivoire.

Afsah, S. (1992) *Extractive Reserves: Economic-Environment Issues and Marketing Strategies for Non-timber Forest Products*. ENVAP and World Bank, Washington, DC.

Agrawal, A. (1997) *Community in Conservation: Beyond Enchantment and Disenchantment*. Conservation & Development Forum, Gainesville, Fla.

Alcorn, J.B. (1989) Process as resource: the traditional agricultural ideology of Bora and Huastec resource management and its implications for research. In: Posey, D.A. and Balée, W. (eds), *Resource Management in Amazonia: Indigenous Folk Strategies*. Advances in Economic Botany Vol. 7. The New York Botanical Garden, Bronx, New York, pp. 63–77.

Allan, W. (1965) *The African Husbandman*. Oliver and Boyd, Edinburgh.

Almeida, M.W.B. (1996) Household extractive economies. In: Ruiz-Pérez, M. and Arnold, J.E.M. (eds), *Current Issues in Non-timber Forest Products Research*. CIFOR, Bogor, Indonesia, pp. 119–141.

Almekinders, C. and de Boef, W. (2000) *Encouraging Diversity: The Conservation and Development of Plant Genetic Resources*. Intermediate Technology Publications, London.

Almekinders, C. and Louwaars, N. (1999) *Farmers' Seed Production: New Approaches and Practices*. Intermediate Technology Publications, London.

Almekinders, C.J.M., Fresco, L.O. and Struik, P.C. (1995) The need to study and manage variation in agro-ecosystems. *Netherlands Journal of Agricultural Science* 43(2): 127–142.

Altieri, M.A. (1995) *Agroecology: The Science of Sustainable Agriculture*. 2nd edn. Westview and Intermediate Technology Publications, Boulder and London.

Altieri, M.A. (1999) The ecological role of biodiversity in agroecosystems. *Agriculture, Ecosystems and Environment* 74(1–3): 19–31.

Amanor, K.S. (1994) *The New Frontier: Farmer Responses to Land Degradation*. INRISD and Zed, Geneva and Atlantic Highlands.

Amanor, K.S. (1995) Indigenous knowledge in space and time. *PLEC News and Views* 5: 26–30.

Anane-Sakyi, C. and Dittoh, S. (2001) Agro-biodiversity conservation: preliminary work on *in situ* conservation and management of indigenous rice varieties in the interior savanna of Ghana. *PLEC News and Views* 17: 31–33.

Anderson, A. and Ioris, E. (1992) The logic of extraction: resource management and resource generation by extractive producers in the estuary. In: Redford, K. and Padoch, C. (eds), *Conservation of Neotropical Forests: Working From Traditional Resource Use*. Columbia University Press, New York, pp. 175–199.

Ardayfio-Schandorf, E. and Awumbila, M. (2000) Gender and agrodiversity in southern Ghana: preliminary findings. *PLEC News and Views* 15: 23–26.

Arnold, J.E.M. (1995) Farming the issues. In: Arnold, J.E.M. and Dewees, P.A. (eds), *Tree Management in Farmer Strategies: Responses to Agricultural Intensification.* Oxford University Press, Oxford, pp. 3–17.

Asante, F. (2002) Adaptation of farmers to climate change: a case study of selected farming communities in the forest-savanna transition zone of southern Ghana. MSc dissertation, Department of Geography and Resource Development, University of Ghana, Legon.

Avery, T.E. and Burkhart, H.E. (1983) *Forest Measurements.* McGraw-Hill, New York.

Barbier, E.B. (1995) The economics of forestry and conservation: economic values and policies. *Commonwealth Forestry Review* 74: 26–34.

Barry, A.K., Fofana, S., Diallo, A. and Boiro, I. (1996) Systèmes de production et changements de l'environnement dans le sous-bassin de Kollangui-Pita. WAPLEC-Guinea, Conakry (mimeo).

Batterbury, S.P.J. and Bebbington, A.J. (1999) Environmental histories. Access to resources and landscape change: an introduction. *Land Degradation and Development* 10: 279–289.

Baudry, J. (1993) Landscape dynamics and farming systems: problems of relating patterns and predicting ecological changes. In: Bunce, R.G.H., Ryszkowski, L. and Paoletti, M.G. (eds), *Landscape Ecology and Agroecosystems.* Lewis Publishers, London, pp. 21–40.

Bebbington, A.J. (1994) Farmers' federations and food systems: organizations for enhancing rural livelihoods. In: Scoones, I. and Thompson, J. (eds), *Beyond Farmer First: Rural Practice, People's Knowledge, Agricultural Research and Extension Practice.* Intermediate Technology Publications, London, pp. 220–224.

Bellon, M.R. (1996) The dynamics of crop infraspecific diversity: a conceptual framework at the farmer level. *Economic Botany* 50(1): 26–39.

Benjasilaraks, M. and Silaraks, S. (1999) *Human Rights of Forest People: Marginal People of Thailand.* Northern Development Foundation, Chiang Mai, Thailand.

Benneh, G. (1972) Systems of agriculture in tropical Africa. *Economic Geography* 48(3): 244–257.

Black, R. and Sessay, M.F. (1997) Forced migration, environmental change and woodfuel issues in the Senegal River Valley. *Environmental Conservation* 24: 251–260.

Blaikie, P. (2000) Development, post-, anti-, and populist: a critical review. *Environment & Planning A* 32(6): 1033–1050.

Blaikie, P. and Brookfield, H. (1987) *Land Degradation and Society.* Methuen, London.

Blaikie, P. and Jeanrenaud, S. (1997) Biodiversity and human welfare. In: Ghimire, K.B. and Pimbert, M.P. (eds), *Social Change and Conservation.* Earthscan, London, pp. 46–70.

Blay, E.T. (2001) Diversity of yams and vegetables in PLEC demonstration sites in southern Ghana, an unpublished PLEC report. Department of Crop Science and Department of Geography and Resource Development, University of Ghana, Legon.

Bodmer, R.E., Penn, J.W., Puertas, P., Moya, P. and Fang, T.G. (1997) Linking conservation and local people through sustainable use of natural resources. In: Freese, C. (ed.), *Harvesting Wild Species: Implications for Biodiversity Conservation.* Johns Hopkins University Press, Baltimore, pp. 315–358.

Boyle, T.J.B. and Sayer, J.A. (1995) Measuring, monitoring and conserving biodiversity in managed tropical forests. *Commonwealth Forestry Review* 74: 20–25.

Brondizio, E.S. (1996) Forest farmers: human and landscape ecology of caboclo populations in the Amazon estuary. Doctoral dissertation, School of Public and Environmental Affairs. Indiana University, Bloomington, IN.

Brondizio, E.S. and Siqueira, A.D. (1997) From extractivists to forest farmers: changing concepts of caboclo agroforestry in the Amazon estuary. *Research in Economic Anthropology* 18: 233–279.

Brookfield, H. (1995) Postscript: the population–environment nexus and PLEC. *Global Environmental Change: Human and Policy Dimensions* 5(4): 381–393.

Brookfield, H. (2001) *Exploring Agrodiversity*. Columbia University Press, New York.

Brookfield, H. and Brown, P. (1963) *Struggle for Land: Agriculture and Group Territories among the Chimbu of the New Guinea Highlands*. Oxford University Press, Melbourne.

Brookfield, H. and Padoch, C. (1994) Appreciating agrodiversity: a look at the dynamism and diversity of indigenous farming practices. *Environment* 36(5): 6–11, 37–45.

Brookfield, H. and Stocking, M. (1999) Agrodiversity: definition, description and design. *Global Environmental Change: Human and Policy Dimensions* 9: 77–80.

Brookfield, H., Stocking, M. and Brookfield, M. (1999) Guidelines on agrodiversity in demonstration site areas. *PLEC News and Views* (Special Issue) 13: 17–31.

Brush, S.B. (1992) Ethnoecology, biodiversity, and modernization in Andean potato agriculture. *Journal of Ethnobiology* 12(2): 161–185.

Brush, S.B. (1995) In situ conservation of landraces in centers of crop diversity. *Crop Science* 35(2): 346–354.

Brush, S.B. (ed.) (1999) *Genes in the Field: Conserving Plant Diversity on Farms*. Lewis, Boca Raton, FL.

Brush, S. (2000) Midterm review: the review report. *PLEC News and Views* 16: 3–7.

Bunch, R. and López, G. (1999) Soil recuperation in Central America: how innovation was sustained after project intervention. In: Hinchcliffe, F., Thompson, J., Pretty, J., Gujit, I. and Shah, P. (eds), *Fertile Ground: The Impacts of Participatory Watershed Management*. Intermediate Technology Publications, London, pp. 32–41.

Cairns, M. (1997) Modification of fallow vegetation to increase swidden productivity: understanding farmer strategies in Southeast Asia. Workshop on Indigenous Strategies for Intensification of Shifting Cultivation in Southeast Asia, Bogor, Indonesia, 23–27 June 1997.

Cairns, M. and Garrity, D.P. (1999) Improving shifting cultivation in Southeast Asia by building on indigenous fallow management strategies. *Agroforestry Systems* 47(1–3): 37–48.

Carney, D. (ed.) (1998) *Sustainable Rural Livelihoods: What Contribution Can We Make?* Department for International Development, London.

Carter, J. (1995) *Alley Farming: Have Resource-poor Farmers Benefited?* Natural Resource Perspectives No. 3. Overseas Development Institute, London.

Christiaensen, L., Tollens, E. and Ezedinma, C. (1995) Development patterns under population pressure: agricultural development and the cassava–livestock interaction in smallholder farming systems in sub-Saharan Africa. *Agricultural Systems* 48: 51–72.

Clay, J. and Clement, C.R. (1993) *Selected Species and Strategies to Enhance Income Generation from Amazonian Forests*. FAO Working Paper. FAO, Rome.

Cleaver, K.M. and Schreiber, G.A. (1994) *Reversing the Spiral: the Population, Agriculture and Environment Nexus in Sub-Saharan Africa*. World Bank, Washington DC.

Clements, F.E. (1916) *Plant Succession: An Analysis of the Development of Vegetation*. Carnegie Institute of Washington Publication 242. Washington, DC.

Coffey, K. (2000) *PLEC Agrodiversity Database Manual. Report for PLEC*. United Nations University (available at www.unu.edu/env/plec).

Collins, W.W. and Qualset, C.O., (eds) (1999) *Biodiversity in Agroecosystems*. CRC Press, Boca Raton, Fla.

Committee on the Role of Alternative Farming Methods in Modern Production Agriculture, Board on Agriculture, N.R.C. (1989) *Alternative Agriculture*. National Academy Press, Washington, DC.

Conelly, W.T. and Chaiken, M.S. (2000) Intensive farming, agro-diversity, and food security under conditions of extreme population pressure in western Kenya. *Human Ecology* 28(1): 19–51.

Conklin, H.C. (1957) *Hanunóo Agriculture in the Philippines*. FAO Forestry Development Paper No.12. FAO, Rome.

Cooper, R.G. (1984) *Resource Scarcity and the Hmong Response: Patterns of Settlement and Economy in Transition*. Singapore University Press, Singapore.

Corbeels, M., Shiferaw, A. and Haile, M. (2000) *Farmers' Knowledge of Soil Fertility and Local Management Strategies in Tigray, Ethiopia*. Managing Africa's Soils No. 10. International Institute for Environment and Development, London.

Critchley, W. (1999) Harnessing traditional knowledge for better land husbandry in Kabale District, Uganda. *Mountain Research and Development* 19: 261–272.

Cromwell, E. (1999) Agriculture, biodiversity and livelihoods: issues and entry points. Available: http://nt1.ids.ac.uk/eldis/agbio.htm (18/11/1999). Overseas Development Institute, London.

Cromwell, E., Cooper, D. and Mulvany, P. (2001) Agriculture, biodiversity and livelihoods: issues and entry points for development agencies. In: Koziell, I. and Saunders, J. (eds), *Living Off Biodiversity*. International Institute for Environment and Development, London, pp. 75–112.

Crosson, P.R. and Stout, A.T. (1983) *Productivity Effects of Cropland Erosion in the United States*. Resources for the Future, Washington DC.

Crowley, E.L. and Carter, S.E. (2000) Agrarian change and the changing relationships between toil and soil in Maragoli, western Kenya (1900–1994). *Human Ecology* 28(3): 383–414.

de Schlippe, P. (1956) *Shifting Cultivation in Africa: The Zande System of Agriculture*. Routledge and Kegan Paul, London.

Denevan, W.M. and Padoch, C. (1987) *Swidden-fallow Agroforestry in the Peruvian Amazon*. Advances in Economic Botany Vol. 5. The New York Botanical Garden, Bronx, New York.

Denevan, W.M. and Treacy, J.M. (1987) Young managed fallows at Brillo Nuevo. In: Denevan, W.M. and Padoch, C. (eds), *Swidden-fallow Agroforestry in the Peruvian Amazon*. Advances in Economic Botany Vol. 5. The New York Botanical Garden, Bronx, New York, pp. 8–46.

Department of Public Welfare (1995) *A Directory for Highland Community in Thailand, 1995*. Department of Public Welfare, Ministry of Labour and Social Welfare, Bangkok.

Derman, W. (1973) *Serfs, Peasants and Socialists: A Former Serf Village in the Republic of Guinea*. University of California Press, Berkeley, Calif.

Dickinson, M.B., Dickinson, J.C. and Putz, F.E. (1996) Natural forest management as a conservation tool in the tropics: divergent views on possibilities and alternatives. *Commonwealth Forestry Review* 75: 309–315.

Dickson, K.B. (1969) *A Historical Geography of Ghana*. Cambridge University Press, Cambridge.

Dufour, D.L. (1990) Use of tropical rainforest by native Amazonians. *BioScience* 40: 652–659.

Dunteman, G. (1989) *Principal Components Analysis*. Sage Publications, New Berry Park, Calif.

Dupire, M. (1970) *L'organisation Sociale des Peul*. Plon, Paris.

Edwards, P.J., Kollmann, J. and Wood, D. (1999) Determinants of agrobio-diversity in the agricultural landscape. In: Wood, D. and Lenné, J.M. (eds),

Agrobiodiversity: Characterization, Utilization and Management. CABI, New York, pp. 183–210.

El Comercio (1996) El camu-camu puede convertirse en el nuevo 'boom' de la selva. Article published in the issue of 9th August, 1996 (Lima, Peru).

Ellis-Jones, J. (1999) Poverty, land care and sustainable livelihoods in hillside and mountain regions. *Mountain Research and Development* 19: 179–190.

Elwell, H.A. (1986) Determination of the erodibility of a subtropical clay soil: a laboratory rainfall simulator experiment. *Journal of Soil Science* 37: 345–350.

Enters, T. (1998) *Methods for the Economic Assessment of the On- and Off-site Impacts of Soil Erosion.* Issues in Sustainable Land Management No.2. International Board for Soil Research and Management, Bangkok.

Expreso (1996) El camu-camu ingresa al mercado de Japon. Article published in the issue of 9th August, 1996 (Lima, Peru).

Fairhead, J. (1993) Representing knowledge: the 'new farmer' in research fashions. In: Pottier, J. (ed.), *Practising Development: Social Science Perspectives.* Routledge, London, New York, pp. 187–204.

Fairhead, J. and Leach, M. (1996) *Misreading the African Landscape: Society and Ecology in a Forest-Savanna Mosaic.* African Studies Series 90. Cambridge University Press, Cambridge.

Falconer, J. (1992) *Non-timber Forest Products in Southern Ghana: The Main Report.* Forestry Department and ODA, Accra, Ghana and London.

FAO (1976) *A Framework for Land Evaluation.* Soils Bulletin 32. FAO, Rome.

FAO (1983) *Guidelines: Land Evaluation for Rainfed Agriculture.* Soils Bulletin 52. FAO, Rome.

FAO (1986) *Food and Fruit-bearing Forest Species: Examples from Latin America.* Forestry Paper 44/3. FAO, Rome.

FAO (1996) *Report on the state of the world's plant genetic resources for food and agriculture prepared for the International Technical Conference on Plant Genetic Resources, Leipzig, Germany, 17–23 June 1996.* FAO, Rome.

FAO (1998) *The State of the World's Plant Genetic Resources for Food and Agriculture.* FAO, Rome.

Feder, G. and O'Mara, G.T. (1981) Farm size and the adoption of green revolution technology. *Economic Development and Cultural Change* 30: 59–76.

Fernandes, E.C.M. and Nair, P.K.R. (1986) An evaluation of the structure and functions of tropical home gardens. *Agricultural Systems* 21(4): 279–310.

Field, M.J. (1943) The agricultural system of the Manya-Krobo of the Gold Coast. *Africa* 14: 54–65.

Fofana, S., Diallo, A., Barry, A.K., Boiro, I., Kane, M.A. and Barry, M.O. (1998) Enlargement and development of tapades by use of compost in the Fouta Djallon, République de Guinée. *PLEC News and Views* 11: 20–22.

Forman, R.T.T. (1995) *Land Mosaics: The Ecology of Landscapes and Regions.* Cambridge University Press, Cambridge.

Fowler, C. and Mooney, P. (1990) *The Threatened Gene: Food, Policies and Loss of Genetic Diversity.* Lutterworth Press, Cambridge.

Franklin, J.F. (1993) Preserving biodiversity: species, ecosystems or landscapes? *Ecological Applications* 3: 202–205.

Garcia, A. (1992) Conserving the species-rich meadows of Europe. *Agriculture, Ecosystems and Environment* 40: 219–232.

Gliessman, S.R. (1998) *Agroecology: Ecological Processes in Sustainable Agriculture.* Ann Arbor Press, Chelsea, MI.

Global Environment Facility (GEF) (2000) *Operational Program No. 13 on Conservation and Sustainable Use of Biological Diversity Important to Agriculture.* Global Environment Facility, Washington DC.

Goulding, M., Smith, N.J.H. and Mahar, D.J. (1996) *Floods of Fortune: Ecology and Economy Along the Amazon.* Columbia University Press, New York.

Gove, J.H., Patil, G.P. and Taillie, C. (1996) Diversity management and comparison with examples. In: Szaro, R.C. and Johnston, D.W. (eds), *Biodiversity in Managed Landscapes: Theory and Practice.* Oxford University Press, New York, pp. 157–175.

Grandstaff, T. (1980) *The Development of Swidden Agriculture (Shifting Cultivation).* Teaching and Research Forum No. 23: A Development and Change reprint. Agricultural Development Council, Bangkok.

Guèye, M.B. (1995) The active method of participatory research and planning (MARP) as a natural resource management tool. In: Stiles, D. (ed.), *Social Aspects of Sustainable Dryland Management.* Wiley, Chichester, pp. 83–92.

Guo Huijun and Padoch, C. (1995) Patterns and management of agroforestry systems in Yunnan: an approach to upland rural development. *Global Environmental Change: Human and Policy Dimensions* 5(4): 273–279.

Guo Huijun and Wu Zhenyu (1998) Agrobiodiversity. In: Guo Huijun and Long Chunlin (eds), *Biodiversity of Yunnan, SW China* [in Chinese]. Yunnan Science and Technology Press, Kunming, China, pp. 112–113.

Guo Huijun, Dao Zhiling and Brookfield, H. (1996) Agrodiversity and biodiversity on the ground and among the people: methodology from Yunnan. *PLEC News and Views* 6: 14–22.

Guo Huijun, Chen Aiguo, Dao Zhiling and Brookfield, H. (1998) Agrodiversity assessment (ABA): definitions, practice and analysis. In: *Agrodiversity: Assessment and Conservation.* Yunnan Science and Technology Press, Kunming, China, pp. 14–22.

Guo Huijun, Li Heng and Dao Zhiling (2000a) Dynamism of socio-economy and biodiversity interaction: a case of Gaoligong Mountains. *Acta Botanica Yunnanica* Supplement XII: 42–51 (in Chinese with English abstract).

Guo Huijun, Padoch, C., Fu Yongneng, Dao Zhiling and Coffey, K. (2000b) Household agrobiodiversity assessment (HH-ABA). *PLEC News and Views* 16: 28–33.

Gupta, A. (1998) *Postcolonial Developments: Agriculture in the Making of Modern India.* Duke University Press, Raleigh, NC.

Gyasi, E.A. (1976) Population pressure and changes in traditional agriculture: case study of farming in Sekesua-Agbelitsom. *Bulletin of the Ghana Geographical Association* 18: 68–87.

Gyasi, E.A. (1992) Emergence of a new oil palm belt in Ghana. *Journal of Economic and Social Geography* 83(1): 39–49.

Gyasi, E.A. (1994) The adaptability of African communal land tenure to economic opportunity: the example of land acquisition for oil palm farming in Ghana. *Africa* 64(3): 391–405.

Gyasi, E.A. (1996a) Landholding and its relationship with biophysical status: case of tenancy farming in Ghana. *PLEC News and Views* 7: 21–25.

Gyasi, E.A. (1996b) The environmental impact and sustainability of plantations in sub-Saharan Africa: Ghana's experiences with oil-palm plantations. In: Benneh, G., Morgan, W. B. and Uitto, J. I. (eds), *Sustaining the Future: Economic, Social, and Environmental Change in Sub-Saharan Africa.* United Nations University Press, Tokyo, pp. 342–357.

Gyasi, E.A. (1997) Background and objectives of the study of production pressure and environmental change in the forest–savanna transition zone. In: Gyasi, E.A. and Uitto, J.I. (eds), *Environment, Biodiversity and Agricultural Change in West Africa: Perspectives from Ghana.* United Nations University Press, Tokyo, pp. 38–42.

Gyasi, E.A. (1998) Land tenure system and traditional concepts of biodiversity conservation. In: Amlalo, D.S., Atsiatsome, L.D. and Fiati, C. (eds), *Biodiversity Conservation: Traditional Knowledge and Modern Concepts,* Proceedings of the

Third UNESCO MAB Regional Seminar on Biosphere Reserves for Biodiversity Conservation and Sustainable Development in Anglophone Africa (BRAAF), Cape Coast, 9-12 March 1997, pp. 16–22.

Gyasi, E.A. (1999) Claim that tenant-farmers do not conserve land resources: counter evidence from a PLEC demonstration site in Ghana. *PLEC News and Views* 12: 10–14.

Gyasi, E.A. and Uitto, J.I. (eds) (1997) *Environment, Biodiversity and Agricultural Change in West Africa: Perspectives from Ghana.* United Nations University Press, Tokyo.

Harris, D.R. and Hillman, G.C. (eds) (1989) *Foraging and Farming. The Evolution of Plant Exploitation.* Unwin Hyman, London.

Heinrich, R. (1995) Environmentally sound harvesting to sustain tropical forests. *Commonwealth Forestry Review* 74: 198–203.

Hill, P. (1963) *The Migrant Cocoa-Farmers of Southern Ghana: A study in Rural Capitalism.* Cambridge University Press, Cambridge.

Hodgkin, T. and Jarvis, D.I. (2001) Strengthening the scientific basis of in situ conservation of agricultural biodiversity on-farm: a global project. *PLEC News and Views* 18: 6–14.

Holling, D.G. (1978) *Adaptive Environmental Assessment and Management.* International Series on Applied Systems Analysis 3. John Wiley, Chichester.

Homma, A. (1992) The dynamics of extraction in Amazonia: an historical perspective. In: Nepstad, D.C. and Schwartzman, S. (eds), *Nontimber Products from Tropical Forests: Evaluation of a Conservation and Development Strategy.* Advances in Economic Botany Vol. 9. New York Botanical Garden, Bronx, New York, pp. 23–31.

Homma, A. (1996) Modernization and technological dualism in the extractive economy in Amazonia. In: Ruiz-Pérez, M. and Arnold, J.E.M. (eds), *Current Issues in Non-timber Forest Products Research.* CIFOR, Bogor, Indonesia, pp. 41–57.

Huber, H. (1993) *The Krobo: Traditional Social and Religious Life of a West African People.* Fribourg University Press, Fribourg, Switzerland.

Hurlbert, S.H. (1984) Pseudoreplication and the design of ecological field experiments. *Ecological Monographs* 54: 187–211.

Huxley, P.A. (1999) *Tropical Agroforestry.* Blackwell Science, Oxford.

Jarvis, D.I. and Hodgkin, T. (eds) (1998) *Strengthening the scientific basis of in situ conservation of agricultural biodiversity on-farm. Options for data collecting and analysis.* Proceedings of a workshop to develop tools and procedures for *in situ* conservation on-farm, 25-29 August 1997, Rome, Italy. International Plant Genetic Resources Institute, Rome.

Jarvis, D., Sthapit, B. and Sears, L. (2000) *Conserving Agricultural Biodiversity in situ: A Scientific Basis for Sustainable Agriculture.* Proceedings of a Workshop, 5–12 July 1999, Pokhara, Nepal. International Plant Genetic Resources Institute, Rome.

Jolliffe, I.T. (1986) *Principal Component Analysis.* Springer-Verlag, New York.

Jongman, R.H., Ter Braak, C.J.F. and Van Tongeren, O.F.R. (1995) *Data Analysis in Community and Landscape Ecology.* Second edition. Cambridge University Press, New York.

Juma, C. (1989) *The Gene Hunters. Biotechnology and the Scramble for Seeds.* African Centre for Technology Studies Research Series No.1. Zed Books, London.

Kaihura, F.B.S., Ndondi, P. and Kemikimba, E. (2000) Agrodiversity assessment in diverse and dynamic small-scale farms in Arumeru, Arusha. *PLEC News and Views* 16: 14–27.

Kalliola, R. and Flores Paitan, S. (1998) Geoecología y desarollo amazónico: estudio integrado en la zona de Iquitos, Peru. Turunyliopisto, Turku, Finland.

Keen, F.G.B. (1972) *Upland Tenure and Land Use in North-Western Thailand.* Siam Communication Ltd for the SEATO Cultural Programme, Bangkok.

Kerr, J. and Sanghi, N.K. (1992) *Indigenous Soil and Water Conservation in India's Semi-Arid Tropics*. Gatekeeper Series No. 34. International Institute for Environment and Development, London.

Keystone Centre (1991) *Keystone International Dialogue Series on Plant Genetic Resources. Oslo Plenary Session. Final Consensus Report: Global Initiative for the Security and Sustainable Use of Plant Genetic Resources*. Genetic Resources Communication Systems, Washington, DC.

Kiome, R.M. and Stocking, M. (1995) Rationality of farmer perception of soil erosion: the effectiveness of soil conservation in semi-arid Kenya. *Global Environmental Change: Human and Policy Dimensions* 5(4): 281–295.

Kirkby, J., O'Keefe, P. and Howorth, C. (2001) Introduction: rethinking environment and development in Africa and Asia. *Land Degradation and Development* 12(3): 195–203.

Koziell, I. (2000) *Diversity not Adversity: Sustaining Livelihoods with Biodiversity*. Rural Livelihoods Department, UK Department for International Development, London.

Kranjac-Berisavljevic, G. and Gandaa, B.Z. (2000) Collection of yam types at Bongnayili-Dugu-Song main demonstration site in northern Ghana. *PLEC News and Views* 15: 27–30.

Krebs, C.J. (1998) *Ecological Methodology*. Addison Wesley Longman, Menlo Park, Calif.

Kunstadter, P., Chapman, E.C. and Sabhasri, S. (1978) *Farmers in the Forests: Economic Development and Marginal Agriculture in Northern Thailand*. University Press of Hawaii for East-West Center, Honolulu.

Kwamena-Poh (1973) *Government and Politics in the Akuapem State: 1730–1850*. Longman, London.

Lampietti, J.A. and Dixon, J.A. (1995) *To See the Forest for the Trees: A Guide to Non-timber Forest Benefits*. Environmental Economics Series No.013. World Bank, Washington, DC.

Leach, M., Mearns, R. and Scoones, I. (1999) Environmental entitlements: dynamics and institutions in community-based natural resource management. *World Development* 27: 225–247.

Li Heng, Guo Huijun and Dao Zhiling (eds) (2000) *Flora of Gaoligong Mountains*. Science Press, Beijing [in Chinese with English preface].

Louette, D. (1999) Traditional management of seed and genetic diversity: What is a landrace? In: Brush, S.B. (ed.), *Genes in the Field: Conserving Plant Diversity on Farms*. Lewis, Boca Raton, Fla, pp. 109–142.

Lugo, A.E. (1995) Management of tropical biodiversity. *Ecological Applications* 5(4): 956–961.

Ma Keping (1994) Measure methods of biodiversity. In: *Principles and Methodologies of Biodiversity Studies*. [in Chinese]. Chinese Science and Technology Press, Beijing, pp. 141–165.

Magrath, W.B. and Arens, P. (1989) *The Costs of Soil Erosion on Java: A Natural Resource Accounting Approach*. Environment Department Working Paper No.18. World Bank, Washington DC.

Magurran, M.E. (1988) *Ecological Diversity and its Measurement*. Princeton University Press, NJ.

Manicad, G. (1996) Biodiversity conservation and development: the collaboration of formal and non-formal institutions. *Biotechnology and Development Monitor* 26: 15–17.

Maxted, N., Ford-Lloyd, B. and Hawkes, J.G. (eds) (1997) *Plant Genetic Conservation: The* in situ *Approach*. Chapman & Hall, London.

Mazzucato, V. and Niemeijer, D. (2000) *Rethinking Soil and Water Conservation in a Changing Society: A Case Study in Eastern Burkina Faso*. Tropical Resource

Management Papers 32. Erosion and Soil and Water Conservation Group, Department of Environmental Sciences, Wageningen University and Research Centre, Wageningen, The Netherlands.

McNeely, J.A. (1988) *Economics and Biological Diversity: Developing and Using Economic Incentives to Conserve Biological Resources*. International Union for the Conservation of Nature and Natural Resources, Gland, Switzerland.

Ministry of Environment, Science and Technology (1998) *Biodiversity Strategy and Action Plan*. Draft for comment. Ministry of Environment, Science and Technology, Accra, Ghana.

Moran, E. (1993) Land use and deforestation in the Brazilian Amazon. *Human Ecology* 21: 1–21.

Mortimore, M. (1998) *Roots in the African Dust: Sustaining the Sub-Saharan Drylands*. Cambridge University Press, Cambridge.

Moss, R.P. (1969) An ecological approach to the study of soils and land use in the forest zone of Nigeria. In: Thomas, M.F. and Whittington, G.W. (eds), *Environment and Land Use in Africa*. Methuen, London, pp. 385–407.

Murnaghan, N. (1999) The impacts of population pressure and resource availability on agrodiversity: a pilot study in Ngiresi Village, Tanzania. MSc dissertation, University of East Anglia, Norwich, UK.

Myers, N., Mittermeier, R.A., Mittermeier, C.G., da Fonseca, G.A.B. and Kent, J. (2000) Biodiversity hotspots for conservation priorities. *Nature* 403(6772): 853–858.

Nabila, J.S. (1997) Population growth and urban demand. In: Gyasi, E.A. and Uitto, J.I. (eds), *Environment, Biodiversity and Agricultural Change in West Africa: Perspectives from Ghana*. United Nations University Press, Tokyo, pp. 76–83.

National Security Council and National Economic and Social Development Board (1993) *A Directory of Highland Communities and Population 1993*. NSC and NESDB, Bangkok.

Netting, R.M. and Stone, M.P. (1996) Agro-diversity on a farming frontier: Kofyar smallholders on the Benue Plains of Central Nigeria. *Africa* 66(1): 52–70.

Niñez, V.K. (1984) *Household Gardens: Theoretical Considerations on an Old Survival Strategy*. Potatoes in Food Systems Research Series Report No 1. International Potato Center, Lima, Peru.

O'Connell, M.A. (1996) Managing biodiversity on private lands. In: Szaro, R.C. and Johnston, D.W. (eds), *Biodiversity in Managed Landscapes: Theory and Practice*. Oxford University Press, New York, pp. 157–175.

Odum, E.P. (1954) *Fundamentals of Ecology*. Saunders, Philadelphia.

Oduro, W. (1993) *WAPLEC-UST Group Project Progress Report and Programmed Activity for 1993*. Final Report. UNU/INRA, Accra, Ghana.

Oduro, W. (1994) *WAPLEC-UST Group Project Progress Report and Programmed Activity for 1994*. Final Report. UNU/INRA, Accra, Ghana.

Oduro, W. (1995) *WAPLEC-UST Group Project Progress Report and Programmed Activity for 1995*. Final Report. UNU/INRA, Accra, Ghana.

Oduro, W. (1997) *WAPLEC-UST Group Project Progress Report and Programmed Activity for 1997*. Final Report. UNU/INRA, Accra, Ghana.

Oduro, W. (1998) *People, Land Management and Environmental Change: WAPLEC-UST Subcluster Project Progress Report*. GEF and UNU, Accra, Ghana.

Office of the Narcotic Crops Control Board and UNFDAC (1983) *A Masterplan for Opium Poppy Cultivating Regions of Thailand*. Office of the Narcotic Crops Control Board and United Nations Fund for Drug Abuse Control, Bangkok.

Okigbo, B.N. (1998) Crops and cropping systems in sub-Saharan Africa. In: Chopra, V.L., Singh, R.B. and Varma, A. (eds), *Crop Productivity and Sustainability Shaping the Future*. Oxford and 1B4 Publishing, Delhi, pp. 588–617.

Oughton, G.A. (1970) *Nikhom Doi Chiangdao: A Resources and Development*

Survey. Report 1: A proposal for development of Phaphuchom village (Meo). TRC Mimeographed. Tribal Research Centre, Chiang Mai.

Oughton, G.A. and Imong, N. (1970) *Nikhom Doi Chiangdao: A Resources and Development Survey*. Report 2: Village location, ethnic composition and economy. TRC Mimeographed. Tribal Research Centre, Chiang Mai.

Owusu-Bennoah, E. and Enu-Kwesi, L. (2000) Investigation into trees that combine effectively with field crops. *PLEC News and Views* 15: 20–22.

Pacey, A. (1990) *Technology in World Civilization*. Blackwell, Oxford.

Padoch, C. (1993) Managing forest fragments and forest gardens in Kalimantan. In: Doyle, J.K. and Schelhas, J. (eds), *Forest Remnants in the Tropical Landscape: Benefits and Policy Applications*. Proceedings of the Symposium presented by the Smithsonian Migratory Bird Center. Smithsonian Institution, Washington, DC.

Padoch, C. and de Jong, W. (1987) Traditional agroforestry practices of native and ribereño farmers in the lowland Peruvian Amazon. In: Goltz, H.L. (ed.), *Agroforestry: Realities, Possibilities and Potentials*. Martinus Nijhoff/ICRAF, Dordrecht, pp. 179–194.

Padoch, C. and de Jong, W. (1991) The house gardens of Santa Rosa: diversity and variability in an Amazonian agricultural system. *Economic Botany* 45(2): 166–175.

Padoch, C. and de Jong, W. (1992) Diversity, variation and change in *Ribereño* agriculture. In: Redford, K. and Padoch, C. (eds), *Conservation of Neotropical Forests: Working From Traditional Resource Use*. Columbia University Press, New York, pp. 158–174.

Padoch, C. and Peters, C. (1993) Managed forest gardens in West Kalimantan, Indonesia. In: Potter, C.S., Cohen, J.I. and Janczewski, D. (eds), *Perspectives on Biodiversity: Case Studies of Genetic Resource Conservation and Development*. American Association for the Advancement of Science Press, Washington, DC., pp. 167–176.

Padoch, C. and Pinedo-Vasquez (1996) Smallholder forest management: looking beyond non-timber forest products. In: Ruiz-Pérez, M. and Arnold, J.E.M. (eds), *Current Issues in Non-timber Forest Products Research*. CIFOR, Bogor, Indonesia, pp. 103–117.

Padoch, C. and Pinedo-Vasquez, M. (1998) Demonstrating PLEC: a diversity of approaches. *PLEC News and Views* 11: 7–9.

Padoch, C. and Pinedo-Vasquez, M. (1999) Farming above the flood in the várzea of Amapá: some preliminary results of the Projeto Várzea. In: Padoch, C., Ayres, J.M., Pinedo-Vasquez, M. and Henderson, A. (eds), *Várzea: Diversity, Development, and Conservation of Amazonia's Whitewater Floodplains*. Advances in Economic Botany Vol. 13. The New York Botanical Garden, Bronx, New York, pp. 345–354.

Padoch, C. and Pinedo-Vasquez, M. Concurrent activities and invisible technologies: an example of timber management in Amazonia. In: Posey, D.A. (ed.), *Human Impacts on the Amazon: The Role of Traditional Ecological Knowledge in Conservation and Development*. Columbia University Press, New York, (in press).

Padoch, C., Harwell, E. and Susanto, A. (1998) Swidden, sawah and in-between: agricultural transformation in Borneo. *Human Ecology* 26(1): 3–20.

Pagiola, S. (1999) *The Global Environmental Benefits of Land Degradation Control on Agricultural Land*. World Bank Environment Paper 16. World Bank, Washington DC.

Peluso, N.L. (1992) *Rich Forests, Poor People: Resource Control and Resistance in Java*. University of California Press, Berkeley, Calif.

Peluso, N.L. and Padoch, C. (1996) Changing resource rights in managed forests of West Kalimantan. In: Padoch, C. and Peluso, N.L. (eds), *Borneo in Transition: People, Forests, Conservation and Development*. Oxford University Press, Kuala Lumpur, pp. 121–136.

Penth, H. (2000) *The Brief History of Lan Na Civilizations of North Thailand.* Silkworm Press, Chiang Mai.

Pereira, V.F.G. (1998) Spatial and temporal analysis of floodplain ecosystems–Amapá, Brazil using geographic information system (GIS) and remote sensing. MSc thesis, University of New Hampshire, Durham, NH.

Perrings, C.A., Mäler, K.-G., Falke, C.S., Holling, C.S. and Jansson, B.-O. (1995) *Biodiversity Conservation: Problems and Policies.* Kluwer Publishers, Dordrecht, The Netherlands.

Peters, C.M. (1996) *The Ecology and Management of Non-timber Forest Resources.* World Bank Technical Paper No. 322. World Bank, Washington, DC.

Peters, C.M. and Hammond, E.J. (1990) Fruits from the flooded forests of Peruvian Amazonia: yield estimates for natural populations of three promising species. In: Prance, G.T. and Balick, M. (eds), *New Directions in the Study of Plants and People: Research Contributions from the Institute of Economic Botany.* Advances in Economic Botany Vol. 8. New York Botanical Garden, Bronx, New York, pp. 159–176.

Phillips, J.M. (1994) Farmer education and farmer efficiency: a meta analysis. *Economic Development and Cultural Change* 43: 149–165.

Pinedo-Panduro, M. (1996) Camu-camu (*Myrciaria dubia* H.B.K McVaugh), una propuesta agroecológica para la Amazonia Baja. *Bosques Amazónicos* 1(2): 7–8.

Pinedo-Vasquez, M. (1995) Human impact on várzea ecosystems in the Napo-Amazon, Peru. Doctoral dissertation, School of Forestry and Environmental Studies, Yale University, New Haven, CT.

Pinedo-Vasquez, M. (1996) Local experts and local leaders: lessons from Amazonia. *PLEC News and Views* 6: 30–32.

Pinedo-Vasquez, M. and Padoch, C. (1996) Managing forest remnants and forest gardens in Peru and Indonesia. In: Schelhas, J. and Greenberg, R. (eds), *Forest Patches in Tropical Landscapes.* Island Press, Washington, DC., pp. 327–342.

Pinedo-Vasquez, M., Zarin, D. and Jipp, P. (1990) Use-values of tree species in a communal forest reserve in northeast Peru. *Conservation Biology* 4: 405–416.

Pinedo-Vasquez, M., Padoch, C., McGrath, D. and Ximenes, T. (2000) Biodiversity as a product of smallholders' strategies for overcoming changes in their natural and social landscapes. *PLEC News and Views* 15: 9–19.

Pinedo-Vasquez, M., Zarin, D., Coffey, K., Padoch, C. and Rabelo, F. (2001) Post-boom logging in Amazonia. *Human Ecology* 29(2): 219–239.

Plumptre, R.A. (1996) Links between utilization, product marketing and forest management in tropical moist forest. *Commonwealth Forestry Review* 75: 316–322.

Prance, G., Balée, W., Boom, B. and Jipp, P. (1987) Quantitative ethnobotany and the case for conservation in Amazonia. *Conservation Biology* 1: 296–310.

Pretty, J.N. (1995) *Regenerating Agriculture: Policies and Practice for Sustainability and Self-reliance.* Earthscan, London.

Primack, R.B. (1998) *Essentials of Conservation Biology.* Sinauer Associates, Sunderland, Mass.

Putz, F.E. and Viana, V. (1996) Biological challenges for certification of tropical timber. *Biotropica* 28(3): 323–330.

Rabelo, M.F. (2000) Niveis da composicion floristica e a estrutura das matas de Mazagao e Ipixuna. Masters thesis, Faculdade de Ciências Agrárias do Pará. Universidade Federal do Pará, Belém, Brasil.

Raffles, H. (1998) Igarapé Guariba: nature, locality, and the logic of Amazonian anthropogenesis. Doctoral dissertation, School of Forestry and Environmental Studies. Yale University, New Haven, CT.

Rakyuthitham, A. (2000) Disaster in a Hmong village: conflict between ethnic groups.

In: *Forests and Land: Resource Management in Watershed*. People Participation Project, Northern Development Foundation, Chiang Mai, pp. 34–45.

Reij, C., Scoones, L. and Toulmin, C., (eds) (1996) *Sustaining the Soil: Indigenous Soil and Water Conservation in Africa*. Earthscan, London.

Republic of Ghana (1990) *Ghana Medium Term Agricultural Development Programme (MTADP): An Agenda for Sustained Agricultural Growth and Development (1991–2000)*. Ministry of Food and Agriculture, Accra, Ghana.

Rerkasem, B., Rerkasem, K. and Shinawatra, B. (1995) A note on mixed cropping by hill-tribe people in Northern Thailand. *PLEC News and Views* 5: 31–32.

Rice, R.E., Gullison, R.E. and Reid, J.W. (1997) Can sustainable management save tropical forests. *Scientific American* 276(4): 44–49.

Richard-Molard, J. (1951) Les densités de population au Fouta-Djallon. In: Pélissier, P., (ed.), Hommage à Jacques Richard-Molard (1913-1951). Special Issue of *La Présence Africaine* 15: 95–106.

Richards, P. (1985) *Indigenous Agricultural Revolution: Ecology and Food Production in West Africa*. Hutchinson, London.

Roberts, M.R. and Gilliam, F.S. (1995) Patterns and mechanisms of plant diversity in forested ecosystems: implications for forest management. *Ecological Applications* 5(4): 969–977.

Robinson, J.G. (1993) The limits to caring: sustainable living and the loss of biodiversity. *Conservation Biology* 7: 20–28.

Rodda, G.H. (1993) How to lie with biodiversity. *Conservation Biology* 7: 1959–1960.

Rounce, N.V. (1949) *The Agriculture of the Cultivation Steppe of the Lake, Western and Central Provinces*. Department of Agriculture, Tanganyika Territory. Longman, Green & Co., Cape Town, South Africa.

Sanchez, P.A. (1995) Science in agroforestry. *Agroforestry Systems* 30: 5–55.

Sanchez, P.A. (1999) Improved fallows come of age in the tropics. *Agroforestry Systems* 47(1 / 3): 3–12.

Sauer, J.D. (1993) *Historical Geography of Crop Plants: A Select Roster*. CRC Press, Boca Raton, Fla.

Scoones, I. (1994) *Living with Uncertainty: New Directions in Pastoral Development in Africa*. Intermediate Technology Publications, London.

Scoones, I. (1996) *Hazards and Opportunities: Farming Livelihoods in Dryland Africa. Lessons from Zimbabwe*. Zed Books, London.

Scoones, I. (1998) *Sustainable Rural Livelihoods: A Framework for Analysis*. Working Paper No. 72. Institute of Development Studies, Brighton, England.

Scoones, I. (1999) New ecology and the social sciences: what prospects for a fruitful engagement? *Annual Review of Anthropology* 28: 479–507.

Scoones, I. (ed.) (2001) *Dynamics and Diversity: Soil Fertility and Farming Livelihoods in Africa*. Earthscan, London.

Scoones, I. and Thompson, J. (1994) Knowledge, power and agriculture: towards a theoretical understanding. In: Scoones, I. and Thompson, J. (eds), *Beyond Farmer First: Rural Practice, People's Knowledge, Agricultural Research and Extension Practice*. Intermediate Technology Publications, London, pp. 16–32.

Scott, J.C. (1998) *Seeing Like a State: How Certain Schemes to Improve the Human Condition Have Failed*. Yale University Press, New Haven, CT.

Sears, R., Padoch, C. and Pinedo-Vasquez, M.(2000) Amazon forestry transformed: integrating knowledge for smallholder timber management in eastern Brazil. (2000). Paper presented at the International Society for Ethno-biology Congress, 22 October 2000, Athens, Georgia, USA.

Seybold, C.A., Herrick, J.E. and Brejda, J.J. (1999) Soil resilience: a fundamental component of soil quality. *Soil Science* 164(4): 224–234.

Smith, R. (1996) Biodiversity won't feed our children. In: Redford, K.H. and Mansour, J.A. (eds), *Traditional Peoples and Biodiversity Conservation in Large Tropical Landscapes*. America Verde Publications, The Nature Conservancy, Arlington, VA, pp. 197–218.

Smith, T.C. (1988) *Native Sources of Japanese Industrialization*, 1750–1920. University of California Press, Berkeley, Calif.

Stakman, E.C., Bradfield, R. and Mengelsdorf, P. (1967) *Campaigns Against Hunger*. Belknap Press, Cambridge, Mass.

Stocking, M.A. (1994) Assessing vegetative cover and management effects. In: Lal, R. (ed.), *Soil Erosion Research Methods*. 2nd edn. Soil and Water Conservation Society, Ankeny, Ia, pp. 210–232.

Stocking, M.A. (1996) Soil erosion. In: Adams, W.M., Goudie, A. and Orme, W (eds), *The Physical Geography of Africa*. Oxford University Press, Oxford, pp. 326–341.

Stocking, M.A. (2000) Biological diversity, land degradation and sustainable rural livelihoods. *Acta Botanica Yunnanica* Supplement XII: 4–17.

Stocking, M.A. and Murnaghan, N. (2001) *Handbook for the Field Assessment of Land Degradation*. Earthscan, London.

Swanson, T.M. (ed.) (1995) *The Economics and Ecology of Biodiversity Decline: The Forces Driving Global Change*. Cambridge University Press, Cambridge.

Tengberg, A. and Stocking, M. (1997) *Erosion-induced Loss in Soil Productivity and its Impacts on Agricultural Production and Food Security*. Land and Water Development Division, FAO, Rome.

Tengberg, A., Ellis-Jones, J., Kiome, R. and Stocking, M. (1998) Applying the concept of agrodiversity to indigenous soil and water conservation practices in eastern Kenya. *Agriculture, Ecosystems and Environment* 70(2–3): 259–272.

Terborgh, J. (1999) *Requiem for Nature*. Island Press, Washington, DC.

Thailand Development Research Institute (1995) *The Economics of Watershed Management: A Case Study of Mae Taeng*. Natural Resources and Environment Program, Thailand Development Research Institute, Bangkok.

Thrupp, L.A. (1998) *Agricultural Biodiversity and Food Security: Predicaments, Policies and Practices*. World Resources Institute, Washington DC.

Tiffen, M., Mortimore, M.J. and Gichuki, F. (1994) *More People, Less Erosion: Environmental Recovery in Kenya*. Wiley, Chichester, UK.

Tisdell, C. (1999) *Biodiversity, Conservation and Sustainable Development: Principles and Practice with Asian Examples*. Edward Elgar, Cheltenham, UK.

Toulmin, C. (1993) *Combating Desertification: Setting the Agenda for a Global Convention*. Drylands Network Programme, Issues Paper 42. International Institute for Environment and Development, London.

Townson, I.M. (1995) *Incomes from Non-timber Forest Products: Patterns of Non-timber Forest Products Enterprise Activity in the Forest Zone of Southern Ghana*. Main report. Forestry Department and Oxford Forestry Institute, Accra, Ghana and London.

Uhl, C., Barreto, P., Verissimo, A., Vidal, E., Amaral, P., Barros, A.C., Souza, C., Johns, J. and Gerwing, J. (1997) Natural resource management in the Brazilian Amazon. *BioScience* 47(3): 160–168.

UNDP Office to Combat Desertification and Drought (2000) *Promoting Farmer Innovation: Harnessing Local Environmental Knowledge in East Africa*. United Nations Development Programme, New York.

United Nations Environment Programme (1992) *Convention on Biological Diversity: Environmental Law and Institutions Programme Activity Centre*. United Nations Environment Programme, Nairobi, Kenya.

United Nations Environment Programme (1994) *The Convention on Biological*

Diversity: Issues of Relevance to Africa, Regional Ministerial Conference on the Convention on Biological Diversity. United Nations Environment Programme, Nairobi, Kenya.

United Nations Environment Programme (1995) *Global Biodiversity Assessment*. Cambridge University Press, Cambridge.

Uphoff, N. (1994) Local organization for supporting people-based agricultural research and extension: lessons from Gal Oya, Sri Lanka. In: Scoones, I. and Thompson, J. (eds), *Beyond Farmer First: Rural Practice, People's Knowledge, Agricultural Research and Extension Practice*. Intermediate Technology Publications, London, pp. 213–220.

van den Breemer, J.P.M., Drijver, C.A. and Venema, L.B. (1995) *Local Resource Management in Africa*. John Wiley, Chichester, UK.

Van Royen, W. (1954) *The Agricultural Resources of the World*. Prentice-Hall, New York.

Vavilov, N.I. (1926) Studies on the origin of cultivated plants. *Bulletin of Applied Botany and Plant Breeding* (Leningrad) 16(1): 1–243.

Visser, B. (1998) *The CBDC programme: community innovation systems and on-farm conservation*. Presentation for the European Symposium on Plant Genetic Resources for Food and Agriculture, 30 June–3 July 1998, Braunschweig, Germany. Available http://www.cbdcprogram.org/BRAU_CBD.htm [9/12/1999].

Vogt, K.A., Larson, C., Gordon, J.C., Vogt, D.J. and Fanzeres, A. (2000) *Forest Certification: Roots, Issues, Challenges and Benefits*. CRC Press, New York.

Walker, A.R. (1975) Two blue Meo communities in Northern Thailand. In: Walker, A.R. (ed.), *Farmers in the Hills: Ethnographic Notes on the Upland Peoples of Northern Thailand*. Penerbit Universiti Sains Malaysia, Pulau Pinang, pp. 73–79.

Wang Xianpu and Yang Zhouhai (1995) Xishuangbanna Region, Yunnan Province, China. In: Davis, S.D., Heywood, V.H. and Hamilton, A.C. (eds), *Centres of Plant Diversity: A Guide and Strategy for their Conservation. Vol.2. Asia, Australasia and the Pacific*. WWF and IUCN, Cambridge, UK.

Whitmore, T.C. (1982) On pattern and process in forests. In: Newman, E.I. (ed.), *The Plant Community as a Working Mechanism*. Special Publication Number 1 of the British Ecological Society Produced as a Tribute to A.S. Watt. Blackwell Scientific Publications, Oxford, pp. 45–59.

Whittaker, R.H. (1965) Dominance and diversity in land plant communities. *Science* 147: 250–260.

Whittaker, R.H. (1975) *Communities and Ecosystems*, 2nd edn. Macmillan, New York.

Wilken, G.C. (1987) *Good Farmers: Traditional Agricultural Resource Management in Mexico and Central America*. University of California Press, Berkeley, Calif.

Wilson, E.O. and Peters, F.M. (eds) (1988) *Biodiversity*. National Academy Press, Washington, DC.

Wood, D. and Lenné, J.M. (eds) (1999) *Agrobiodiversity: Characterization, Utilization and Management*. CABI and Oxford University Press, New York.

World Bank (1992) *World Development Report 1992: Development and the Environment*. Oxford University Press, New York.

World Bank (1996) *Desertification: Implementing the Convention*, 2nd edn. The World Bank, Washington DC.

Xishuangbanna Tropical Botanical Garden and Ethnobotanical Department of Kunming Institute of Botany (1996) *Name List of Higher Plants in Xishuangbanna, Yunnan*. Nationality Press, Kunming, China.

Zarin, D.J. (1995) Diversity measurement for PLEC clusters. *PLEC News and Views* 4: 11–21.

Zarin, D.J., Guo Huijun and Enu-Kwesi, L. (1999) Methods for the assessment of plant species diversity in complex agricultural landscapes: guidelines for data collection and analysis from the PLEC Biodiversity Advisory Group (PLEC-BAG). *PLEC News and Views* 13: 3–16.

Zhang Aoluo and Wu Sugong (1996) Introduction. In: Zhang Aoluo and Wu Sugong (eds), *Floristic Characteristics and Diversity of East Asian Plants*. Springer-Verlag, New York.

Zimmerer, K.S. (1996) *Changing Fortunes: Biodiversity and Peasant Livelihood in the Peruvian Andes*. University of California Press, Berkeley, Calif.

Zimmerer, K.S. and Young, K.R. (1998) Introduction: the geographical nature of land-scape change. In: Zimmerer, K.S. and Young, K.R. (eds), *Nature's Geography: New Lessons for Conservation in Developing Countries*. University of Wisconsin Press, Madison, Wis, pp. 3–34.

Zinke, P., Sabhasri, S. and Kunstadter, P. (1978) Soil fertility aspects of the Lua forest fallow system of shifting cultivation. In: Kunstadter, P., Chapman, E.C. and Sabhasri, S. (eds), *Farmers in the Forests: Economic Development and Marginal Agriculture in Northern Thailand*. University Press of Hawaii for East-West Center, Honolulu, pp. 134–159.

Zohary, D. and Hopf, M. (1993) *Domestication of Plants in the Old World. The Origin and Spread of Cultivated Plants in West Asia, Europe, and the Nile Valley*. Clarendon Press, Oxford, UK.

Index